ESSENTIAL
MENTAL MATHS
PRACTICE TESTS

Graham Newman

Nelson

Thomas Nelson and Sons Ltd
Nelson House
Mayfield Road
Walton-on-Thames
Surrey KT12 5PL
United Kingdom

I(T)P® Thomas Nelson is an International Thomson Company

I(T)P® is used under licence

First published by Thomas Nelson and Sons Ltd 1997

ISBN 0–17–431500–7
NPN 9 8 7 6 5 4 3

Typeset in 9.5pt Stone Sans by ⫟ Tek-Art, Croydon, Surrey
Printed in China

Edited by First Class

Contents

Introduction   iv

Part 1: Mental Maths Practice Tests   1

Section A   2
Section B   22
Section C   42
Section D   62
Section E   82

Part 2: Answer Grids   103

Section A   104
Section B   124
Section C   144
Section D   164
Section E   184

Introduction

This book offers a range of mental tests for general use in schools. The book is divided into two parts:

- Part 1: Mental Maths Practice Tests and Answers
- Part 2: Answer Grids

Both parts are divided into five sections, labelled A to E. In each section of Part 1, there are 40 tests of 25 questions, accompanied by the answers. In each section of Part 2, there are 40 matching answer grids which can be photocopied for use in the classroom.

The questions in each section approximately relate to National Curriculum levels as follows:

Section	A	B	C	D	E
Levels	2–3	3–4	4–5	5–6	6–8

Sections B to E correspond to Books 1 to 4 in the *Essential Skills in Maths* series, also published by Thomas Nelson.

A recommended approach to the selection of a suitable test for a group of students would be to estimate the level at which the group is working, then select a test at which that level is the highest. Thus, students working at or about level 4 could be given a test from Section B (level 3–4). These tests include consolidation at the lower level 3, which also provide preparation for the harder questions at level 4.

Computational skills relate to Number within the National Curriculum, although contexts are drawn from other aspects of the National Curriculum, which require knowledge of facts which relate to other Attainment Targets. The following metric–Imperial conversions are used:

- 1 inch ≈ 2.5 centimetres
- 1 foot ≈ 30 centimetres
- 1 yard ≈ 1 metre
- 5 miles ≈ 8 kilometres

- 1 metre ≈ 39 inches
- 1 kilogram ≈ 2.2 lbs
- 1 litre ≈ 1.75 pints
- 4.5 litres ≈ 1 gallon

Part I: Mental Maths Practice Tests

Each test comprises 25 questions which test students' mental arithmetic, estimation skills and knowledge of mathematical terms. Consecutive tests follow a common format of similarly phrased questions, allowing an opportunity for familiarisation with the wording of the questions that may be asked, and consolidation of the techniques needed to arrive at a correct answer. Within each test, the questions fall into three different types.

- The first type are short questions, which students might normally be expected to complete in about five seconds. These questions involve simple, single calculations or mental recall.

- The second type of question includes those which students might normally be expected to answer in about ten seconds. These questions involve short step calculations.

- Students would not normally be expected to complete the third type of question in under fifteen seconds. These questions are more complex and require more computational skills.

It is suggested that each question is read out twice, and then the students be allowed the time (5, 10 or 15 seconds) specified for that type of question. During this time, the students should work out the answer in their heads and write down only their final answer.

When reading out the test, it is advisable to pause after each section, and to remind the students how much time they will have to answer after the question has been read out twice. No question should be repeated after its second reading.

Each test should take no more than 15–20 minutes to administer.

Answers are provided beside the questions and advice about marking is given in the introduction to the Answer Grid Section on page 103.

A1

1 Alana is twenty-nine years old. What will be her age on her next birthday? — 30

2 What is six less than ten? — 4

3 Write in figures: the number three hundred and twenty-three. — 323

4 Alan has three stamps, Barbara has five, and Craig has two. How many stamps have they, altogether? — 10

5 Write down the largest number in this list: eighty, thirty-three, sixty-eight, eighty-two and sixty-four. — 82

6 From twenty-six, take away fourteen. — 12

7 Write in figures: the number one thousand, two hundred. — 1200

8 Twelve boys stay for football practice; sixteen girls stay for netball practice; how many stay after school altogether? — 28

9 There are four apples in a packet. How many apples are there, altogether, in three packets? — 12

10 After seven days, how many days of a fifteen-day holiday remain? — 8

11 How many minutes are there, between five past one and half past one? — 25 min

12 The scores on two dice are six and five. What is the total score on the two dice? — 11

13 Julie grows twenty cucumbers. She uses thirteen of them. How many cucumbers remain? — 7

14 Louise has four pounds to spend at the cinema. She gives half to her friend Kelly. How much will Kelly have? — £2

15 What is the greatest number of two-pence coins you could have for seventeen pence? — 8

16 What is sixty, take away twenty-four? — 36

17 Find the total of three ten-pence coins and three five-pence coins. — 45p

18 What is ten more than five hundred and ninety-eight? — 608

19 A clock runs five minutes fast. When the clock shows half past one, what is the correct time? — 1.25*

20 Add three hundred and twenty to sixty-four. — 384

21 Look at the numbers on your answer sheet. Draw a circle around each even number. — 38, 56

22 Look at the numbers on your answer sheet. Write these numbers in order of size, starting with the largest. — 32, 31, 23, 19

23 Add forty-eight to fifty-eight. — 106

24 A tin of beans costs twenty-five pence. What is the cost of three tins of beans? — 75p

25 Write down the year that is exactly four hundred years after fifteen hundred and thirty. — 1930

A2

1 Write down the smallest number in this list: twenty-eight, forty-seven, fifty-five, twenty-six, thirty-three. — 26

2 What is one less than sixty? — 59

3 What is the total of four pence, three pence, and five pence? — 12p

4 Write in figures: the number four hundred and thirty-five. — 435

5 How many individual shoes are there, in two pairs of shoes? — 4

6 A train sets off with sixteen passengers. Another fourteen passengers board the train at its first stop. How many passengers are there now, on the train? — 30

7 Six felt-tip pens are shared equally between two pupils. How many does each one receive? — 3

8 Write in figures: the number one thousand and eighty. — 1080

9 What is the greatest number of ten-pence coins you could have for forty pence? — 4

10 A DJ has four hundred and four records, but loses seven of them. How many records will he now have? — 397

11 What is the greatest number of two-pence coins you could have for nine pence? — 4

12 Martin has eight sweets. He eats half of them. How many are left? — 4

13 Malika buys eighteen envelopes. She uses twelve of them. How many are left? — 6

14 How many minutes are there, between quarter past one and quarter to two? — 30 min

15 Penny scores thirteen and six at darts. What is her total score? — 19

16 Add twenty-five and twenty-seven. — 52

17 A man has two hundred and ninety-nine pounds in the bank. He has another ten pounds which he also puts in the bank. How much will he now have, in the bank? — £309

18 What is the time, twenty minutes after quarter to ten? — 10.05*

19 How much change do you get from two pounds, if you spend one pound and seventy-five pence? — 25p

20 From forty-four, take twenty-three. — 21

21 Look at the numbers on your answer sheet. Write them in order of size, starting with the largest. — 813, 754, 633, 572

22 Look at the numbers on your answer sheet. Draw a circle around each of the even numbers. — 36, 24

23 An antique cupboard has a value of three thousand, seven hundred and ten pounds. It is sold for five hundred pounds more. For what price was it sold? — £4210

24 Add thirty-seven and seventy-seven. — 114

25 Multiply two hundred and thirty-seven by two. — 474

A3

1. A pet shop has seventy-one fish. Two are sold. How many fish remain? — 69
2. Write in figures: the number three hundred and eighty. — 380
3. Write down the smallest number in this list: sixty-eight, forty-six, seventy-two, seventy-nine, forty-three. — 43
4. Add together one, three, and seven. — 11
5. A box contains three golf balls. How many golf balls are there, altogether, in three boxes? — 9
6. A CD costs fourteen pounds. What would be the cost of buying two CDs? — £28
7. Write in figures: the number four thousand, three hundred and twenty. — 4320
8. How many hours are there, between half past ten in the morning and half past four in the afternoon? — 6 hrs
9. How many tens are there, in forty? — 4
10. Kazi has sixteen rabbits. She sells eleven of them. How many does she have left? — 5

11. What is the greatest number of five-pence coins you could have for sixteen pence? — 3
12. David has a fifteen-metre roll of wire. He uses six metres from the roll. What length is left? — 9 m
13. An apple pie is cut into quarters. How many pieces will there be? — 4
14. Sohaib has eleven tokens and collects six more. How many will he now have? — 17
15. You have a one-pound coin and buy a loaf of bread for eighty-eight pence. What change should you receive? — 12p
16. Add forty-nine and twenty-four. — 73
17. What is the time, ten minutes after five to four? — 4.05*
18. What is fifty-two, take away thirty-eight? — 14
19. Anita has driven three hundred and ninety-eight kilometres. She drives a further thirty kilometres. What total distance has she now driven? — 428 km
20. What is thirty less than three hundred? — 270

21. Look at the numbers on your answer sheet. Draw a circle around each of the odd numbers. — 79, 57, 41
22. Look at the numbers on your answer sheet. Write them in order of size, starting with the smallest. — 176, 267, 269, 354
23. Add eighty-seven and thirty-four. — 121
24. Write down the year that is exactly five hundred years before fourteen hundred and sixty. — 960
25. There are fifty-four people on a coach. Twenty-eight people get off. How many people are left on the coach? — 26

A4

1. A plant of height thirty-nine centimetres grows by another one centimetre. What is its new height? — 40 cm
2. Write down the largest number in this list: twenty-nine, thirty-seven, forty, seventeen, thirty-six. — 40
3. There are four pairs of socks on a washing line. How many individual socks are on the line? — 8
4. Write in figures: the number two hundred and ninety-nine. — 299
5. Add together four, six, and three. — 13
6. What is the greatest number of five-pence coins you could have for thirty-five pence? — 7
7. Write in figures: the number one thousand, nine hundred. — 1900
8. What is the total of eight pence, nine pence, and seven pence? — 24p
9. You have a part-time job which pays ten pounds a week. How much will you earn over four weeks? — £40
10. There are three hundred and one pupils in a school. Five leave. How many remain at the school? — 296

11. What is the greatest number of ten-pence coins you could have for fifteen pence? — 1
12. How many minutes are there, between quarter past two and ten to three? — 35 min
13. Mark flies for fourteen hours in a plane and, later, flies for a further eight hours. What is the total time, he has spent in a plane? — 22 hrs
14. You have fifteen pounds at the start of the week. You spend seven pounds and, then, five pounds. How much money will you have left? — £3
15. Look at the line on your answer sheet. Put an X halfway along the line. — (bisector)
16. What is the time, ten minutes before ten past ten? — 10.00*
17. Add twenty-nine and thirty-two. — 61
18. A temperature starts at fifty-seven degrees and falls by fourteen degrees. What does the temperature fall to? — 43°
19. William has saved nine pounds and thirty-five pence, but wants ten pounds. How much more does he need? — 65p
20. What is ten more than three hundred and ninety-eight? — 408

21. Look at the numbers on your answer sheet. Write them in order of size, starting with the largest. — 864, 823, 730, 699
22. Look at the numbers on your answer sheet. Draw a circle around each odd number. — 13, 15, 25
23. Add thirty-seven and seventy-six. — 113
24. Write down the year that is three hundred years after the year fifteen hundred and fifty. — 1850
25. A book costs three pounds and fifty pence. Jeremy saves fifty pence per week. How many weeks will Jeremy have to save until he can buy the book? — 7 wks

1 Write down the largest number in this list: thirty-eight, seventeen, thirty-nine, twenty-nine, twenty-eight.   **39**

2 There are twenty-eight pupils in a class. Two more pupils join the class. How many are there, now?   **30**

3 Write in figures: the number four hundred.   **400**

4 A cat was born in nineteen ninety-five. How old was the cat in nineteen ninety-seven?   **2 yrs**

5 There are five tapes in a packet. How many tapes are there, altogether, in two packets?   **10**

6 Write in figures: the number two thousand and fifty.   **2050**

7 How many hours are there, between half past six in the morning and half past eight in the evening?   **14 hrs**

8 A woman has sailed fifty-two days of a sixty-day journey. How many days has she before the journey is over?   **8 days**

9 What is four times three?   **12**

10 What is four hundred and thirty-nine, to the nearest hundred?   **400**

---

11 Adam scores a total of nineteen points in two games. He scored seven points in his first game. How many points did he score in his second game?   **12**

12 What is the total cost of two twelvepenny stamps?   **24p**

13 Jomo buys a packet of five washers. He already has eighteen washers. How many does he now have, in total?   **23**

14 What is the greatest number of fivepenny stamps you could have for nineteen pence?   **3**

15 There are eight apples in a bag. One quarter of them are bruised. How many apples are bruised?   **2**

16 Add twenty-five and seventeen.   **42**

17 How much change do you get from ten pounds, if you spend four pounds and forty pence?   **£5.60**

18 Take twelve away from forty-six.   **34**

19 A rabbit weighed one hundred and seventy-five grams. It has now doubled in weight. What is its new weight?   **350 g**

20 A watch runs ten minutes slow. When the watch shows twenty to eleven, what is the correct time?   **10.50***

---

21 Look at the numbers on your answer sheet. Draw a circle around each odd number.   **17, 13**

22 Look at the numbers on your answer sheet. Write them in order of size, starting with the smallest.   **45, 54, 78, 106**

23 A piece of wood is two thousand, eight hundred and thirty millimetres long. Another piece is seven hundred millimetres long. What is the total length of the two pieces of wood?   **3530 mm**

24 Add forty-eight and fifty-four.   **102**

25 How much is three glasses costing two pounds and ninety-nine pence each?   **£8.97**

1 A man is one year less than fifty years old. How old is the man now?   **49**

2 Write in figures: the number two hundred and seventy.   **270**

3 Add together the numbers: two, eight, and three.   **13**

4 Write down the smallest number in this list: eighty-four, sixty-six, eighty-three, forty-eight, sixty-two.   **48**

5 Five members of a nine-piece band are girls. How many are boys?   **4**

6 A journey to Limerick in Ireland takes fourteen hours. The return journey takes sixteen hours. What was the total travelling time, taken for the whole journey?   **30 hrs**

7 What is five times two?   **10**

8 Add together eight, seven, nine, and two.   **26**

9 Write in figures: the number one thousand, one hundred and one.   **1101**

10 What is the greatest number of two-pence coins you could have for twenty pence?   **10**

---

11 How many minutes are there, between ten past eleven and twenty-five past eleven.   **15 min**

12 There are ten market stalls in a high street, but only half are being used. How many stalls are empty?   **5**

13 What is the greatest number of five-pence coins you could have for twenty-one pence?   **4**

14 The temperature at dawn is thirteen degrees. By noon, the temperature has risen by eleven degrees. What is the temperature at noon?   **24°**

15 A journey takes twenty-one hours. After twelve hours, how many hours of travelling remain?   **9 hrs**

16 Add thirty-seven and fourteen.   **51**

17 How many seconds are there, in ten minutes?   **600 sec**

18 Add forty-five and fifty-five.   **100**

19 One bottle of shampoo contains one hundred and twenty-five millilitres. How much will two bottles contain?   **250 ml**

20 What is ten more than five hundred and ninety-eight?   **608**

---

21 Look at the numbers on your answer sheet. Draw a circle around each of the even numbers.   **50, 46**

22 Look at the numbers on your answer sheet. Write them in order of size, starting with the smallest.   **245, 254, 258, 265**

23 Add seventy-five and seventy-six   **151**

24 A path is forty-seven metres long. The length is reduced by twenty-nine metres. What is the new length of the path?   **18 m**

25 Multiply two hundred and thirty-four by two.   **468**

1 Lisa has a collection of sixty-nine rocks. She is given one more rock. How many will she now have? 70

2 Write down the smallest number in this list: twenty-seven, forty, twenty-eight, thirty-six, twenty-one. 21

3 Write in figures: the number eight hundred and thirty. 830

4 Andrew was five years old in nineteen ninety-eight. In which year was he born? 1993

5 Add together these numbers: five, four, two. 11

6 What is five less than four hundred and four? 399

7 Write in figures: the number three thousand and ninety-five. 3095

8 Brian has twenty-three pounds saved. He spends seventeen pounds. How much savings will he have left? £6

9 Fifteen counters are divided equally between three pupils. How many counters will each pupil receive? 5

10 What is fifty-seven, to the nearest ten? 60

11 What is the greatest number of ten-pence coins you could have for nineteen pence? 1

12 A cake is cut in half. How many pieces of cake will there be? 2

13 Ann cycles for eight kilometres, then, for six kilometres and, then, for five kilometres. How far did she cycle, altogether? 19 km

14 A tank contains twenty-six litres of water. Sixteen litres of water are drained from the tank. How many litres of water remain? 10 litres

15 How many minutes are there, between twenty-five past four and ten to five? 25 min

16 Find the total of two ten-pence coins and five five-pence coins. 45p

17 A video recorder priced at two hundred and ninety-nine pounds is reduced by sixty-nine pounds. What is its reduced price? £230

18 From thirty, take away twelve. 18

19 Work out forty-seven, take away thirty-one. 16

20 A watch is ten minutes fast. When the watch is showing half past eight, what is the correct time? 8.20*

21 Look at the numbers on your answer sheet. Write them in order of size, starting with the largest. 861, 854, 531, 165

22 Look at the numbers on your answer sheet. Draw a circle around each of the odd numbers. 43, 53

23 Add eighty-nine and twenty-nine. 118

24 There are forty-one plastic boxes in each crate. How many plastic boxes will there be in five crates? 205

25 Add eight hundred and ten pounds to one thousand, four hundred and thirty-seven pounds. £2247

1 Write down the largest number in this list: twenty-seven, forty-six, eighteen, thirty-nine, twenty-six. 46

2 Write in figures: the number seven hundred and ninety-nine. 799

3 There are fifty-one trees in a wood, but two trees are cut down. How many trees are left? 49

4 What is two multiplied by three? 6

5 You have eight pounds, but spend six pounds. How many pounds are left? £2

6 Write in figures: the number four thousand, three hundred and two. 4302

7 Find the total weight of seventeen grams, eight grams, and seven grams. 32 g

8 A car has four tyres. How many tyres are needed for two cars? 8

9 From sixty, take away fifty-three. 7

10 What is seven less than eight hundred and two? 795

11 David buys sixteen kilograms of potatoes to add to the six kilograms he has already. How many kilograms will he now have, in total? 22 kg

12 There are twenty-one butterflies on a bush. Six fly away. How many are left? 15

13 What is the greatest number of two-pence coins you could have for thirteen pence? 6

14 A log is four metres in length. One quarter is cut off. What is the length of the piece that has been cut off? 1 m

15 How many ten-pence coins are there, in one pound? 10

16 Add nineteen, twenty-one, and twenty-three. 63

17 What is the time, fifteen minutes after ten to eleven? 11.05*

18 Add thirty-three and fifty-seven. 90

19 How much change do you get from ten pounds, if you spend three pounds and fifty pence? £6.50

20 Take seventeen away from thirty-two. 15

21 Look at the numbers on your answer sheet. Write them in order of size, starting with the smallest. 64, 68, 78, 86

22 Look at the numbers on your answer sheet. Draw a circle around each even number. 10, 30, 52

23 A car is driven for two journeys, each of which are two hundred and fifty-four kilometres in length. What is the total distance travelled? 508 km

24 Add seventy-two and sixty-seven. 139

25 Bill has three thousand, eight hundred pounds in his bank account. He pays in another five hundred pounds. How much will he now have? 4300

1. A man bought a new car in nineteen ninety-three. How old was the car in nineteen ninety-five? — 2 yrs
2. Write down the smallest number in this list: sixty-three, seventy-two, forty-three, forty-five, seventy-three. — 43
3. What is three less than nine? — 6
4. Forty mustard seeds are planted, but one does not grow. How many seeds grow into mustard plants? — 39
5. Write in figures: the number eight hundred and three. — 803
6. Amber cycles eight kilometres before lunch and nine kilometres afterwards. How far did she cycle, altogether? — 17 km
7. Twenty-five crayons are shared equally between five pupils. How many crayons does each pupil receive? — 5
8. Take twenty-seven away from forty. — 13
9. Write in figures: the number one thousand and eighty-nine. — 1089
10. Write forty-three, to the nearest ten. — 40

11. How many minutes are there, between ten o'clock and twenty-five past ten? — 25 min
12. Eddy wants to buy a new iron costing twenty pounds. He has only fourteen pounds. How much more money does he need? — £6
13. What is the greatest number of five-pence coins you could have for seventeen pence? — 3
14. A tank contains fifteen litres of oil. Another eleven litres of oil is poured in. How many litres of oil will there now be, in the tank? — 26 litres
15. Look at the shape on your answer sheet. Draw a line on the shape to divide it into halves. — (bisector)
16. Add forty-five and forty-five. — 90
17. A hi-fi unit is reduced from two hundred and fifty pounds to one hundred and ninety-pounds. By how much has it been reduced? — £60
18. Add twenty-two and forty-nine. — 71
19. Find the total of five ten-pence coins and four five-pence coins. — 70p
20. Add together these numbers: eighteen, twenty, twenty-two. — 60

21. Look at the numbers on your answer sheet. Draw a circle around each even number. — 44, 18, 52
22. Look at the numbers on your answer sheet. Write them in order of size, starting with the smallest. — 145, 165, 172, 232
23. What is fifty-three, take away twenty-seven? — 26
24. A toy costs three pounds and twenty-five pence. Kate saves fifty pence each week. How many weeks will Kate have to save until she can buy the toy? — 7 wks
25. Add fifty-nine and forty-four. — 103

1. A garden pond is home to eighteen frogs. Two more frogs arrive one summer. How many frogs are there, now? — 20
2. Write in figures: the number nine hundred and fifteen. — 915
3. What is four multiplied by two? — 8
4. Write down the smallest number in this list: thirty-nine, forty-six, thirty-eight, seventeen, forty-nine. — 17
5. What is two less than six? — 4
6. There are twenty people waiting for a lift which can carry no more than ten people. How many journeys will the lift have to make? — 2
7. Write in figures: the number three thousand and eighty-nine. — 3089
8. Add together eight, six, nine, and seven. — 30
9. There are eighteen girls and twelve boys in a class. How many are in the class, altogether? — 30
10. A temperature of thirty degrees falls by eighteen degrees. What does the temperature fall to? — 12°

11. What is the greatest number of twopenny stamps you could buy with seven pence? — 3
12. How many minutes are there, in a quarter of an hour? — 15 min
13. Betty buys sixteen kilograms of apples at a market but finds that five kilograms need to be thrown away. How much does she have left? — 11 kg
14. Gemma scores eight marks in her first test, seven marks on her second test, and nine marks on her third test. What is her total number of marks for the three tests? — 24
15. A class has thirty pupils. Twenty are girls. How many boys are there? — 10
16. Add seventy-two and fifty-six. — 128
17. Take twenty-three away from thirty-nine. — 16
18. Claire needs twelve pounds for a CD, but has saved only eleven pounds and forty-five pence. How much more does she need? — 55p
19. Add eighty-two and eighteen. — 100
20. There are three hundred and ninety-five passengers on a ferry. Ten more passengers board the ferry. What is the total number of passengers? — 405

21. Look at the numbers on your answer sheet. Draw a circle around each of the even numbers. — 22, 66
22. Look at the numbers on your answer sheet. Write them in order of size, starting with the largest. — 235, 193, 147, 89
23. There are fifty-four calendars in a box. How many calendars will there be, altogether, in five such boxes? — 270
24. Add thirty-nine and sixty-nine. — 108
25. What is the total cost of three magazines costing one pound and ninety-nine pence each? — £5.97

1 Write down the smallest number in this list: thirty-seven, twenty-six, twenty-nine, eighteen, twenty-three. **18**

2 Take two away from eighty-one. **79**

3 Write in figures: the number eight hundred and thirty-eight. **838**

4 There are three wheels on a tricycle. How many wheels are there, altogether, on two tricycles? **6**

5 Derek has seven pounds. He spends four pounds. How much has he left? **£3**

6 How many hours are there, between twenty to two in the afternoon and twenty to eleven in the evening? **9 hrs**

7 Write in figures: the number three thousand and one. **3001**

8 From forty, take thirty-six away. **4**

9 What is four multiplied by five? **20**

10 A train ticket costs five pounds. What would be the cost of eight tickets? **£40**

11 Fran works six hours on Monday, seven hours on Tuesday, and six hours on Wednesday. What is the total hours worked over the three days? **19 hrs**

12 What is the greatest number of ten-pence coins you could have for twenty-five pence? **2**

13 Sam has twenty-two pages of typing to do and has typed thirteen pages already. How many pages are left to do? **9**

14 A pizza has been cut into quarters. How many pieces of pizza are there? **4**

15 How many minutes are there, between five past three and half past three? **25 min**

16 What is ten more than six hundred and ninety-eight? **708**

17 What is the time, twenty minutes before five past one? **12.45***

18 What is forty-two, take away twenty-eight? **14**

19 A set of videos costing forty-four pounds is halved in price. What is the new price? **£22**

20 Add fourteen and twenty-nine. **43**

21 Look at the numbers on your answer sheet. Write them in order of size, starting with the largest. **832, 802, 799, 789**

22 Look at the numbers on your answer sheet. Draw a circle around each of the odd numbers. **17, 57, 21**

23 There are eighty-nine people in a train carriage. There are forty-two people in the next carriage. How many people are in the two carriages, altogether? **131**

24 Kath wants five pounds and seventy-five pence to buy a present, but has saved only three pounds and fifty pence. How much more does she need? **£2.25**

25 Multiply one hundred and thirty-two by two. **264**

1 Write down the largest number in this list: fifty-nine, seventy-eight, forty-eight, seventy-one, thirty-four. **78**

2 What is one less than thirty? **29**

3 Jomo works five hours on Saturday and five hours on Sunday. What is the total number of hours he has worked? **10 hrs**

4 Write in figures: the number eight hundred and ten. **810**

5 Katy sends out party invitations to ten people. Eight people can come to the party. How many people cannot come? **2**

6 Add together these numbers: four, six, nine, eight. **27**

7 Write the number three hundred and thirty-three, to the nearest hundred. **300**

8 A farmer takes twenty-six hours to harvest one field and twelve hours to harvest another. What is the total time, spent harvesting? **38 hrs**

9 Fifty counters are shared equally between five pupils. How many does each pupil receive? **10**

10 Write in figures: the number three thousand, five hundred and five. **3505**

11 A two-metre length of ribbon is cut in half. How long is each piece of ribbon? **1 m**

12 How many minutes are there, between twenty to eleven and eleven o'clock? **20 min**

13 The total distance around a triangle is nineteen centimetres. One side measures five centimetres and another side measures nine centimetres. What is the length of the third side? **5 cm**

14 What is the greatest number of five-pence coins you could have for twelve pence? **2**

15 Kate has thirteen pounds in her piggy bank and eight pounds in her wallet. What is the total amount she has? **£21**

16 Work out forty, take away twenty-six. **14**

17 A trailer contains thirty-eight kilograms of grain. Fourteen kilograms is added. How much grain will there now be, in the trailer? **52 kg**

18 What is ten more than one hundred and ninety-nine? **209**

19 How many seconds are there, in twenty minutes? **1200 sec**

20 Add thirty-three and sixty-seven. **100**

21 Look at the numbers on your answer sheet. Draw a circle around each odd number. **51, 93, 47**

22 Look at the numbers on your answer sheet. Write them in order of size, starting with the smallest. **46, 87, 91, 127**

23 What is fifty-three, take away thirty-seven? **16**

24 Five people playing a game share fifty-three cards equally. How many cards are left over? **3**

25 Peter has sixty-seven pence in his piggy bank. Paul has thirty-eight pence. How much do they have together? **105p**

A13

1. Write down the largest number in this list: thirty-eight, twenty-nine, thirty-six, sixty-three, fifty-one.   63
2. What is one more than nineteen?   20
3. There are nine actors in a play. Seven are women. How many are men?   2
4. Write in figures: the number two hundred and seventy.   270
5. Jennifer bought a house in nineteen ninety-one. By nineteen ninety-four, how many years had Jennifer been in the house?   3 yrs
6. Add together these numbers: five, seven, three, eight.   23
7. It takes Abdul ten hours to put a model plane together. How long would it take him to put seven model planes together?   70 hrs
8. Write in figures: the number four thousand and fifty-five.   4055
9. What is four times five?   20
10. There are fifty people on a bus. Thirty-seven get off. How many remain?   13

11. What is the greatest number of threepenny stamps you could have for ten pence?   3
12. Angelo starts his holiday on the sixth of July and finishes on the twenty-first of July. How many days of holiday did he have?   15 days
13. How many minutes are there, between twenty-five minutes past eight and ten to nine?   25 min
14. Rachel scores thirteen marks in her first test and eight marks in her second test. How many marks did she receive, altogether?   21
15. Look at the line on your answer sheet. Put an X halfway along the line.   (bisector)
16. Add forty-seven and twenty-six.   73
17. Find the total of four ten-pence coins and four five-pence coins.   60p
18. From thirty-three, subtract sixteen.   17
19. How much change should I receive from ten pounds, if I have spent five pounds and fifty pence?   £4.50
20. Add together these numbers: eighteen, nineteen, twenty.   57

21. Look at the numbers on your answer sheet. Write them in order of size, starting with the smallest.   223, 541, 655, 684
22. Look at the numbers on your answer sheet. Draw a circle around each of the even numbers.   24, 66
23. A magazine costs two pounds and eighty pence. Richard saves fifty pence each week. How many weeks will Richard have to save before he can buy the magazine?   6 wks
24. Add seventy-three and forty-six.   119
25. Write down the year that is exactly three hundred years after seventeen hundred and fifty.   2050

A14

1. Write down the smallest number in this list: forty-two, twenty-nine, forty-five, twenty-eight, forty.   28
2. Write in figures: the number five hundred and thirty-five.   535
3. What is one less than eighty?   79
4. There are eight people sat at a lunch table. Four finish their lunch. How many are left eating?   4
5. There are four apples in a packet. How many apples will there be in three packets?   12
6. Write in figures: the number one thousand, seven hundred.   1700
7. There are seventy-three pupils in a school. Sixteen new pupils join the school. How many will there now be?   89
8. Write the number five hundred and sixty-two, to the nearest hundred.   600
9. Add together eighteen millimetres, seven millimetres, and six millimetres.   31 mm
10. What are three threes?   9

11. The first three even numbers are two, four, and six. What is the total of these numbers?   12
12. There are eight people in a bus. Half of them are men. How many are women?   4
13. Richard scores fourteen marks in a test and seven marks in the next test. What is his total score for the two tests?   21
14. What is the greatest number of fourpenny stamps you could buy for ten pence?   2
15. Angela arrives at a bus stop at fifteen minutes past ten. She waits nine minutes for the bus. At what time, does the bus arrive?   10.24*
16. From sixty, take thirty-four away.   26
17. There are two hundred and twenty-five packets of cereal on a supermarket shelf. There are one hundred and three packets in the store room. How many packets are there, altogether?   328
18. What is ten more than six hundred and ninety-three?   703
19. What is the total cost of five pens at fifteen pence each?   75p
20. Add twenty-one and thirty-six.   57

21. Look at the numbers on your answer sheet. Write them in order of size, starting with the largest.   499, 495, 465, 456
22. Look at the numbers on your answer sheet. Draw a circle around each even number.   54, 10, 32
23. Add eighty-three and forty-seven.   130
24. What is the total cost of three books each costing three pounds and ninety-nine pence?   £11.97
25. In a competition, thirty-six people have each won a five-pound prize. What is the total prize money?   £180

1 Ali has seventy-eight points in a game. He gains two more points. What is his total number of points? **80**

2 Write down the largest number in this list: thirty-nine, twenty-seven, forty-two, thirty-five, eleven. **42**

3 Write in figures: the number one hundred and fifty-three. **153**

4 What is four less than six? **2**

5 There are two Easter eggs in a box. How many Easter eggs are there, in four boxes? **8**

6 What is sixteen milligrams, take away eight milligrams, take away five milligrams? **3 mg**

7 There were thirty-five biscuits in a packet. Thirteen have been eaten. How many are left? **22**

8 Write in figures: the number three thousand and four. **3004**

9 What is two times five? **10**

10 Add together eight, seven, nine, and three. **27**

11 What is the greatest number of five-pence coins you could have for twenty-one pence? **4**

12 Look at the line shown on your answer sheet. Divide this line into quarters. **(bisectors)**

13 How many minutes are there, between three o'clock and ten to four? **50 min**

14 Teresa has cards with the numbers seven, four, three, and two. What is the total value of the numbers on her cards? **16**

15 A teacher opens a pack of twenty-four new books and gives out sixteen of them. How many books remain? **8**

16 One tool weighs one hundred and seventy-five grams. What will be the weight of two such tools? **350 g**

17 Find the total of five ten-pence coins and five five-pence coins. **75p**

18 What is seventy-three and twenty-seven? **100**

19 A guinea pig weighed one hundred and fifty grams. It has now doubled in weight. What is its new weight? **300 g**

20 Add thirty-eight and fifty-four. **92**

21 Look at the numbers on your answer sheet. Draw a circle around each odd number. **41, 57**

22 Look at the numbers on your answer sheet. Write them in order of size, starting with the largest. **523, 510, 421, 411**

23 There are eighty-three questions in a test. Martin has thirty-six wrong. How many has he answered correctly? **47**

24 What is two hundred and thirty-five, multiplied by two? **470**

25 On a farm, there are forty-two chickens in one shed, and sixty-eight chickens in another shed. How many chickens are there, altogether? **110**

1 Write down the largest number in this list: sixty-three, thirty-five, forty, fifty-one, sixty-one. **63**

2 What is two less than sixty-one? **59**

3 There are two pairs of socks in a drawer. How many individual socks is this? **4**

4 Write in figures: the number two hundred and forty-nine. **249**

5 There are ten pens in a box. Three will not work. How many pens are left that will work? **7**

6 Fifteen pencils are divided equally between three people. How many pencils will each person receive? **5**

7 Write in figures: the number one thousand, two hundred and three. **1203**

8 There are forty people on a bus. Seven get off and four get on. How many are there now, on the bus? **37**

9 There are seven hundred and six pupils at a school. At the end of a term, eight pupils leave. How many will there now be, at the school? **698**

10 Add twenty-three and fifty-two. **75**

11 What is the greatest number of threepenny chewing gums you could buy with fourteen pence? **4**

12 There are sixteen girls and twelve boys in a class. What is the total number of pupils in the class? **28**

13 How many minutes are there, between ten past eight and half past eight? **20 min**

14 There are four wheels on a car. Half of the wheels need new brakes. How many wheels need new brakes? **2**

15 There are sixteen trees in a garden. Seven of the trees are apple trees. How many of the trees are not apples trees? **9**

16 Add twenty-eight and forty-seven. **75**

17 Find the total of two ten-pence coins and two two-pence coins. **24p**

18 In a sale, the price of a hat is reduced from nine pounds and ninety-nine pence to five pounds and forty-nine pence. How much is the reduction? **£4.50**

19 What is seventy-two, take away fifty-three? **19**

20 Add these numbers: eighteen, twenty-one, thirty. **69**

21 Look at the numbers on your answer sheet. Draw a circle around each odd number. **55, 71**

22 Look at the numbers on your answer sheet. Write them in order of size, starting with the smallest. **261, 465, 564, 736**

23 What is forty-five, added to eighty-seven? **132**

24 A blank video tape costs two pounds and thirty-five pence. Sean saves fifty pence each week. How many weeks will Sean have to save before he can buy the video tape? **5 wks**

25 Multiply twenty-five by three. **75**

1 Write down the largest number in this list: sixty-eight, eighty-eight, eighty-seven, seventy-six, thirty-one.   88

2 A boy bought a new book in nineteen ninety-two. How old was the book in nineteen ninety-six?   4 yrs

3 Write seven hundred and nine in figures.   709

4 What is six less than nine?   3

5 A suitcase costs eighty-eight pounds. Its price is increased by two pounds. What will be its new price?   £90

6 What is the greatest number of five-pence coins you can have for thirty-eight pence?   7

7 There are fifty musicians in an orchestra. Thirty-three are female. How many are male?   17

8 What is two times four?   8

9 Write in figures: the number two thousand and one.   2001

10 Write the number seven hundred and forty-five, to the nearest one hundred.   700

11 Wendy has twenty-eight mints in her bag. She gives away twelve. How many mints does she have now?   16

12 What is the greatest number of fourpenny stamps you could buy for fifteen pence?   3

13 Sheena was born in nineteen seventy-seven. How old was she on her birthday in nineteen ninety-seven?   20 yrs

14 Look at the shape on your answer sheet. Draw lines to divide the shape into quarters.

15 A family spends seven days in France and eight days in Spain. What is the total number of days they spent on holiday?   15 days

16 Take one hundred and twenty-two from one hundred and fifty-six.   34

17 Adam paid forty-five pounds for a meal and twenty-six pounds for a theatre ticket. What was the total cost of the meal and the ticket?   £71

18 What is thirty-four, take away eighteen?   16

19 One section of fencing measures ninety-five centimetres. What is the total length of ten sections of fencing?   950 cm

20 Add together these numbers: thirteen, fourteen, twenty-one.   48

21 Look at the numbers on your answer sheet. Write them in order of size, starting with the largest.   661, 356, 254, 84

22 Look at the numbers on your answer sheet. Draw a circle around each odd number.   37, 31, 67

23 What is the year exactly four hundred years after eighteen forty?   2240

24 Add seventy-eight and fifty-five.   133

25 A glass has a weight of twenty-five grams. What would be the total weight of five glasses?   125 g

1 Write down the smallest number in this list: forty-two, twenty-five, twenty-eight, thirty, forty-seven.   25

2 Write in figures: the number four hundred and seventy.   470

3 What is one more than fifty-nine?   60

4 What is two multiplied by three?   6

5 Bill has seven chocolates. He eats two. How many has he left?   5

6 How many hours are there, between quarter past two in the afternoon and quarter past five in the evening?   3 hrs

7 Prize money of one hundred pounds is shared equally between ten people. How much does each person receive?   £10

8 Write in figures: the number nine thousand and eighty-seven.   9087

9 The contents of a box weigh sixty-six grams. The box weighs twenty-two grams. What is the total weight of the box and its contents?   88 g

10 What is nineteen less than twenty-three?   4

11 James buys sixteen packets of crisps. He already has six packets. How many does he now have, altogether?   22

12 How many minutes are there, between half past ten and eleven o'clock?   30 min

13 A school kitchen has twenty-five litres of milk. They use eighteen litres. How many litres of milk remain?   7 litres

14 What is the greatest number of five-pence coins you could have for fourteen pence?   2

15 Look at the line on your answer sheet. Put an X halfway along the line.   (bisector)

16 What is thirty-four less than sixty?   26

17 Sally has saved six pounds and thirty-five pence. She is given another two pounds and forty pence. How much will she now have, altogether?   £8.75

18 Find the total of these numbers: twenty-one, twenty, nineteen.   60

19 A watch is five minutes fast. When the watch shows half past ten, what is the correct time?   10.25*

20 Find the total of two ten-pence coins, two five-pence coins, and two two-pence coins.   34p

21 Look at the numbers on your answer sheet. Draw a circle around each even number.   20, 32, 14

22 Look at the numbers on your answer sheet. Write them in order of size, starting with the largest.   401, 400, 398, 389

23 What is eighty-three, take away twenty-six?   57

24 What is the total cost of three cassette tapes each costing two pounds and ninety-nine pence?   £8.97

25 Add fifty-four and forty-six.   100

1 In one packet, there are five biscuits. How many biscuits will there be in two packets? 10
2 What is one less than twenty? 19
3 Write down the largest number in this list: seventy-five, twenty-nine, thirty-eight, eighty-one, seventy-one. 81
4 Write in figures: the number one hundred and fifteen. 115
5 Barry is running a six-kilometre race. He has run three kilometres. How much further has he yet to run? 3 km
6 What is twelve, added to twenty-eight? 40
7 Write in figures: the number two thousand, five hundred and seven. 2507
8 To ten metres, add fourteen metres and, then, take away three metres. 21 m
9 Add together four, three, seven, and two. 16
10 Sixteen cards are shared equally between four people. How many cards will each person receive? 4

11 What is the greatest number of four-penny sweets you could buy with eleven pence? 2
12 There are eight people on a minibus. One quarter of those on the bus are men. How many of them are women? 6
13 What is fourteen less than forty? 26
14 How many minutes are there, between quarter past seven and twenty to eight? 25 min
15 Patrick has twenty-one items to wash. He puts sixteen in the first wash. How many items remain? 5
16 Add these numbers: twenty-three, twenty, twenty-five. 68
17 A fork has a handle of length eight centimetres and a prong of length nine centimetres. What is the total length of the fork? 17 cm
18 What is thirty-four, added to fifty-eight? 92
19 What is the cost of five packets of mints at twenty-nine pence each? £1.45
20 What is ten more than four hundred and ninety-eight? 508

21 Look at the numbers on your answer sheet. Write them in order of size, starting with the largest. 243, 231, 123, 112
22 Look at the numbers on your answer sheet. Draw a circle around each even number. 58, 40, 32
23 Add thirty-four and seventy-six. 110
24 A pole has a length of two thousand, eight hundred and fifty millimetres. The pole is extended by a further seven hundred millimetres. What length will it become? 3550 mm
25 Multiply forty-four by five. 220

1 Write down the largest number in this list: fifty-eight, fifty-one, eighty-three, fifty-four, thirty-three. 83
2 Write in figures: the number one hundred and twenty-three. 123
3 What is two less than thirty-one? 29
4 What is two times three? 6
5 There are ten bananas in a bag. Five have been eaten. How many remain? 5
6 Write in figures: the number three thousand, one hundred and one. 3101
7 There are eighty-six people staying in a hotel. Six leave, but eight new people arrive. How many people are there now, in the hotel? 88
8 What is forty, take away twenty-eight? 12
9 What is the greatest number of two-pence coins you could have for twenty pence? 10
10 Kevin has three ten-pound notes. What is the total amount of money he has? £30

11 What is the greatest number of five-pence coins you could have for nineteen pence? 3
12 Jason is paid nine pounds, four pounds, and seven pounds for some work he does. What is the total amount he has been paid? £20
13 Kate has twenty-eight litres of petrol. She uses nineteen litres. How much petrol is left? 9 litres
14 A six-metre length of wire is cut into half. What is the length of each piece? 3 m
15 How many minutes are there, between quarter past three and twenty-five to four? 20 min
16 What is ten more than five hundred and ninety-seven? 607
17 There are fifty-two weeks in the year. Twenty-three weeks have gone. How many more weeks are left? 29 wks
18 Add sixty-six and thirty-three. 99
19 Damian cuts a hose in three equal lengths of eight metres each. What was the original length of the hose? 24 m
20 Add these numbers: twenty-five, eleven, twenty-one. 57

21 Look at the numbers on your answer sheet. Draw a circle around each odd number. 31, 47
22 Look at the numbers on your answer sheet. Write them in order of size, starting with the smallest. 864, 870, 887, 895
23 Add forty-eight and thirty-seven. 85
24 A cupboard costs three hundred and twenty-seven pounds. What would be the total cost of two cupboards? £654
25 Five people share forty-eight cakes equally. How many cakes are left over? 3

A21

1 There are five numbers in a list.
Three are even.
How many are odd? — 2

2 What is two less than forty? — 38

3 Write down the smallest number in this list: forty, forty-one, sixty-three, forty-five, twenty-four. — 24

4 Write in figures: the number eight hundred and three. — 803

5 A CD was released in nineteen ninety-five. How old was the CD in nineteen ninety-eight? — 3 yrs

6 Write in figures: the number two thousand and ninety-nine. — 2099

7 Add together twelve centimetres, seven centimetres, and five centimetres. — 24 cm

8 What is two multiplied by two? — 4

9 There are twenty-two dogs and fourteen cats at a kennel. How many pets are there, altogether? — 36

10 What is twenty-three, take away sixteen? — 7

11 I take seventeen records to a party but can only find nine records to bring home. How many records have I lost? — 8

12 What is seven, added to eighteen? — 25

13 A restaurant has thirteen tables in one room and six in another room. How many tables has it, altogether? — 19

14 What is the greatest number of threepenny stamps you could buy with eight pence? — 2

15 A temperature falls from fifty-three degrees to twenty-nine degrees. By how many degrees has the temperature fallen? — 24°

16 Look at the shape on your answer sheet. Draw a line on the shape to divide it into halves.

17 Add twenty to the number four hundred and eighty-three. — 503

18 What is the total of the numbers twenty-five, eighteen, and twenty-six? — 69

19 A shop reduced the cost of a packet of tuna fish from one pound, ninety-nine pence to one pound, forty-nine pence. By how many pence has it been reduced? — 50p

20 Add twenty-seven to sixty-three. — 90

21 Look at the numbers on your answer sheet. Draw a circle around each of the even numbers. — 18, 32

22 Look at the numbers on your answer sheet. Write them in order of size, starting with the smallest. — 326, 360, 386, 465

23 Write down the year that is exactly three hundred years before two thousand and ten. — 1710

24 Add thirty-eight and seventy-six. — 114

25 A watch costs four pounds and ten pence. Abdul saves fifty pence each week. How many weeks will Abdul have to save before he can buy the watch? — 9 wks

A22

1 You had ten pence but have spent seven pence. How much money have you left? — 3p

2 What is one more than sixty-nine? — 70

3 Write down the smallest number in this list: sixty-eight, eighty-four, forty-five, twenty-eight, thirty-seven. — 28

4 What is the total value of four two-pence coins? — 8p

5 Write in figures: the number three hundred and forty. — 340

6 In a game, there are fifty cards. These are shared equally between ten people. How many cards should each person receive? — 5

7 Add fifteen to thirty-two. — 47

8 What is five hundred and fifty-five, to the nearest hundred? — 600

9 Write in figures: the number two thousand, four hundred and one. — 2401

10 What will be the date, exactly one week after the fifteenth of January? — 22nd Jan

11 What is the greatest number of four-penny stamps you could buy for five pence? — 1

12 Jane has five litres of petrol in her car. She buys twelve litres. How many litres of petrol does she now have? — 17 litres

13 I have seventeen pages left to read in a book. I read nine pages. How many pages will I then have left to read? — 8

14 How many minutes are there, between ten past nine and twenty-five to ten? — 25 min

15 Add forty-six and twenty-five. — 71

16 A pork pie is divided into quarters. How many pieces of pie will there be? — 4

17 Take twelve away from twenty-eight. — 16

18 A clock is five minutes slow. When the clock shows half past three, what is the correct time? — 3.35*

19 What is the total cost of three tickets at four pounds each and two tickets at five pounds each? — £22

20 What is ten more than seven hundred and ninety-one? — 801

21 Look at the numbers on your answer sheet. Draw a circle around each odd number. — 87, 61, 49

22 Look at the numbers on your answer sheet. Write them in order of size, starting with the smallest. — 210, 561, 653, 820

23 From forty-four, take away twenty-seven. — 17

24 A man pays thirty-two five-pound notes in to the bank. What is the total amount of money paid in? — £160

25 Add eighty-seven and twenty-nine. — 116

1 Write down the smallest number in this list: fifty-four, fifty-five, sixty-three, nineteen, twenty-four.   19

2 Write in figures: the number four hundred and fifty-five.   455

3 What is one less than ninety?   89

4 Mary is nine years old. She is four years older than her brother. How old is her brother?   5 yrs

5 What is three times two?   6

6 A shop has sixty packets of crisps. Forty-six have been sold. How many remain?   14

7 What are three tens?   30

8 How many hours are there, between ten past ten in the morning and ten past eight in the evening?   10 hrs

9 Write in figures: the number one thousand and five.   1005

10 From fifteen kilometres, take five kilometres and, then, add another fifteen kilometres.   25 km

11 There are ten counters in a bag. Half of them are blue. How many blue counters are there, in the bag?   5

12 Tracey has nineteen pet mice. She gives twelve of them to good homes. How many mice does she have left?   7

13 What is the greatest number of five-pence coins you could have for eighteen pence?   3

14 What is thirty-three, added to seventeen?   50

15 How many minutes are there, between ten past seven and twenty to eight?   30 min

16 What is the date, eleven days after the eighth of March?   19th Mar

17 A kitten weighed two hundred and fifty grams. It has now doubled in weight. What is its new weight?   500 g

18 What is ninety, take away forty-six?   44

19 The rainfall for last year was sixty-three centimetres. This year the rainfall is fourteen centimetres less. What is the rainfall this year?   49 cm

20 What is the time, twenty minutes after five to eleven?   11.15*

21 Look at the numbers on your answer sheet. Write these numbers in order of size, starting with the largest.   79, 63, 34, 31

22 Look at the numbers on your answer sheet. Draw a circle around each of the odd numbers.   65, 51

23 Add seventy-three and twenty-seven.   100

24 A glass contains three thousand, six hundred and fifty millilitres of water. Sixty millilitres of water are added. How much water is there now, in the glass?   3710 ml

25 What is three hundred and sixteen, multiplied by two?   632

1 Write down the largest number in this list: eighty-seven, fifty-four, sixty-one, sixty-three, sixty-five.   87

2 Write in figures: the number two hundred and ninety.   290

3 What is two more than twenty-nine?   31

4 What is five multiplied by two?   10

5 There are eight pigeons in a shed. Five fly away. How many are left?   3

6 There are seats for four people in each car on an amusement ride. What is the greatest number of people that could fit into three cars?   12

7 Add fourteen and twenty-four.   38

8 The scores on two dice are six and five. What is the total score?   11

9 Write in figures: the number one thousand, two hundred.   1200

10 A box contains two hundred eggs. Three are cracked and are removed. How many are left?   197

11 Postcards cost six pence each. How many postcards can you buy for fifteen pence?   2

12 Gary scores eight marks on one test and five marks on another. What is his total score for the two tests?   13

13 What is half of two pounds?   £1

14 How many days of my fifteen-day holiday remain after seven days?   8 days

15 What is twenty-two less than forty?   18

16 What is the total value of an eightpenny stamp and a fivepenny stamp?   13p

17 Add these numbers: twenty-five, twenty, fourteen.   59

18 Alex is exactly twice the height of his brother, who is seventy centimetres tall. How tall is Alex?   140 cm

19 Add thirty-three and fifty-four.   87

20 From sixty-four, take thirty-seven away.   27

21 Look at the numbers on your answer sheet. Write them in order of size, starting with the smallest.   642, 645, 685, 851

22 Look at the numbers on your answer sheet. Draw a circle around each of the even numbers.   80, 64

23 There are seventy-four parents in a hall for a meeting. Thirty-six parents then arrive. How many parents are there now, in the hall?   110

24 What is thirty-seven less than fifty-two?   15

25 There are five books in a packet. What is the total number of books in forty-three packets?   215

A25

Answers

1 Write down the smallest number in this list: eighty-one, seventy, thirty-two, fifty-six, eighty-five. **32**
2 What is ten, take away four? **6**
3 Write in figures: the number four hundred and forty-eight. **448**
4 A coat is priced at sixty-nine pounds. The price is increased by two pounds. What is the new price? **£71**
5 A vase was made in nineteen ninety-four. How old was the vase by the year nineteen ninety-seven? **3 yrs**
6 What is the greatest number of five-pence coins you could have for fifty pence? **10**
7 There are twenty-seven pupils in a class. Eighteen are boys. How many are girls? **9**
8 Write in figures: the number eight thousand and thirty-seven. **8037**
9 A prize of ten pounds is shared equally by two people. How much will each one receive? **£5**
10 Add five, eight, seven, and eight. **28**

11 Simon has eleven fishing floats. Karen has twelve fishing floats. How many do they have between them? **23**
12 What is the greatest number of ten-pence coins you could have for twenty-two pence? **2**
13 A builder needs twenty-five patio slabs. He has eighteen. How many more does he need? **7**
14 How many minutes are there, between five past three and twenty to four? **35 min**
15 Look at the line shown on your answer sheet. Divide this line into quarters. **(bisectors)**
16 Add fifty-nine and twelve. **71**
17 A watch is ten minutes fast. When the watch shows twenty to six, what is the correct time? **5.30***
18 What is twenty more than four hundred and eighty-five? **505**
19 How much change do you get from five pounds, if you spend four pounds and thirty-five pence? **65p**
20 What is thirty-seven, take away twenty-nine? **8**

21 Look at the numbers on your answer sheet. Draw a circle around each even number. **76, 40**
22 Look at the numbers on your answer sheet. Write them in order of size, starting with the smallest. **199, 264, 265, 358**
23 From sixty-four, take thirty-six. **28**
24 What is the cost of four small toys at one pound, ninety-nine pence each? **£7.96**
25 Add fifty-six and fifty-six. **112**

A26

Answers

1 Write down the largest number in this list: ten, sixty, forty-nine, thirty-two, fifty. **60**
2 What is two less than forty-one? **39**
3 During the seven days of a week, on three days it rained. On how many days was it dry? **4**
4 How many litres are there, altogether, in three cartons of milk, each of which contains two litres? **6 litres**
5 Write in figures: the number seven hundred and thirty-two. **732**
6 What is two multiplied by ten? **20**
7 What is the number twenty-six, to the nearest ten? **30**
8 Add together fifteen centimetres, six centimetres, and twelve centimetres. **33 cm**
9 What is the total of sixteen pence and twenty-two pence? **38p**
10 Write in figures: the number eight thousand and eight. **8008**

11 Stephen has a case which weighs twelve kilograms, and a bag which weighs three kilograms. What is the total weight of his luggage? **15 kg**
12 What is the greatest number of two-pence coins you could get for fifteen pence? **7**
13 What is the date, twelve days after the fifth of November? **17th Nov**
14 Zak was born in nineteen eighty-two. How old was he in nineteen ninety-seven? **15 yrs**
15 What is ten more than four hundred and ninety-four? **504**
16 An apple is cut in half. How many pieces of apple are there? **2**
17 Add ten, twenty-three, and twenty-five. **58**
18 How many seconds are there, in ten minutes? **600 sec**
19 What is eighteen less than thirty? **12**
20 Louise has gained sixty-eight marks in one test and thirty-two marks in another test. How many marks has Louise gained, altogether? **100**

21 Look at the numbers on your answer sheet. Write them in order of size, starting with the largest. **688, 687, 678, 668**
22 Look at the numbers on your answer sheet. Draw a circle around each even number. **86, 54, 42**
23 What is fifty-six, multiplied by five? **280**
24 A piece of pipe is seven thousand, four hundred and eighty millimetres long. The pipe is lengthened by another six hundred millimetres. What will be the new length of the pipe? **8080 mm**
25 Add fifty-nine and forty-seven. **106**

A27

1 Write down the largest number in this list: forty-nine, forty-eight, thirty-eight, fifty-eight, fifty.
    58

2 What is one more than forty-nine?
    50

3 Write in figures: the number eight hundred and forty.
    840

4 What change would you get from a five-pound note, if you are spending two pounds?
    £3

5 What is the total value of two five-pence coins?
    10p

6 Write in figures: the number two thousand and forty-two.
    2042

7 In a class of thirty pupils, eighteen have pets at home. How many pupils do not have a pet?
    12

8 Add twenty-five and fifty-two.
    77

9 Find the total value of twenty coins, if each coin is a two-pence coin.
    40p

10 Eight biscuits are shared equally between four children. How many biscuits does each child receive?
    2

11 Sarah has fourteen pounds in her bag and eight pounds in her pocket. How much money has she, altogether?
    £22

12 What is twenty, take away nine?
    11

13 How many minutes are there, between twenty-five to seven and five to seven?
    20 min

14 What is the greatest number of fourpenny sweets you could buy for fourteen pence?
    3

15 Look at the line on your answer sheet. Divide this line into quarters.
    (bisectors)

16 Take nineteen away from thirty-seven.
    18

17 What is the time, ten minutes before five past three?
    2.55*

18 There are twenty-two players in a school football competition and sixty-eight spectators. How many people are there, altogether?
    90

19 Add twenty-eight and forty-one.
    69

20 What is ten more than four hundred and ninety-six?
    506

21 Look at the numbers on your answer sheet. Write them in order of size, starting with the smallest.
    221, 335, 633, 654

22 Look at the numbers on your answer sheet. Draw a circle around each odd number.
    27, 51, 45

23 Add eighty-eight and twenty-one.
    109

24 A cassette tape costs five pounds and forty pence. Sheena saves fifty pence each week. How many weeks will Sheena have to save before she can buy the tape?
    11 wks

25 Take twenty-eight away from forty-five.
    17

A28

1 Write down the smallest number from this list: sixty-four, sixty-nine, eighty, seventy-five, sixty-seven.
    64

2 What is one more than seventy-nine?
    80

3 Write in figures: the number seven hundred and eighty.
    780

4 What is four multiplied by two?
    8

5 There are eight shirts on a line. Three are white. How many are coloured?
    5

6 Write in figures: the number one thousand and ninety-five.
    1095

7 Sixteen sweets are shared equally between four children. How many sweets does each child receive?
    4

8 Take seven kilograms from fifteen kilograms but, then, add nine kilograms. What is the final weight?
    17 kg

9 How many hours are there, between eight o'clock in the morning and midday?
    4 hrs

10 Write the number three hundred and eighty-seven, to the nearest hundred.
    400

11 Ian uses six exercise books during the first half of the year and eight more in the second half of the year. How many books does he use during the whole year?
    14

12 How many minutes are there, between twenty to three and five to three?
    15 min

13 There are eight counters in a bag. One quarter of the counters are red. How many counters are coloured red?
    2

14 There are nineteen light bulbs in a house. Five of them do not work. How many bulbs do work?
    14

15 What is the greatest number of threepenny stamps you could buy for eleven pence?
    3

16 What is sixty-eight, take away thirty-eight?
    30

17 A box weighs six kilograms and its contents weigh thirty-five kilograms. What is the total weight?
    41 kg

18 Add thirty-five and thirty-eight.
    73

19 A clock is twenty minutes slow. When the clock shows half past four, what is the correct time?
    4.50*

20 Add these numbers: twenty, twenty-three, sixteen.
    59

21 Look at the numbers on your answer sheet. Draw a circle around each odd number.
    35, 27

22 Look at the numbers on your answer sheet. Write them in order of size, starting with the largest.
    381, 297, 285, 172

23 Multiply two hundred and thirty-one by two.
    462

24 Bill has to make a journey of sixty-three kilometres. He has already driven forty-nine kilometres. How far has he yet to go?
    14 km

25 Add thirty-eight and sixty-eight.
    106

1 Shelly has a carrot, seven centimetres in length. She cuts off five centimetres. What length of carrot remains? 2 cm
2 What is two less than fifty? 48
3 Write down the smallest number in this list: twenty-three, thirty-three, nineteen, thirty-eight, forty-five. 19
4 Write in figures: the number nine hundred and ninety. 990
5 A bike was bought in nineteen ninety-one. How old was the bike in nineteen ninety-five? 4 yrs
6 A camera tripod has three legs. What is the total number of legs of five camera tripods? 15
7 From fifty, take away thirty-six. 14
8 A gold watch costs seventy-two pounds and its strap costs another twenty-seven pounds. What is the total cost of the watch and strap? £99
9 Write in figures: the number eight thousand and ninety-nine. 8099
10 Add together the numbers: two, seven, six, and seven. 22
11 A puppy weighed three hundred and fifty grams. It has now doubled in weight. What is the new weight? 700 g
12 What is the date, fourteen days after the seventh of July? 21st July
13 Andrew was nine years old in nineteen ninety-five. In which year was he born? 1986
14 County scored twelve goals in the August and seven goals in September. How many goals is this, altogether? 19
15 What is the greatest number of ten-pence coins you could have for twelve pence? 1
16 What is sixteen less than thirty-eight? 22
17 Add thirty-eight and forty-two. 80
18 There are two hundred and seventy-five millilitres left in a bottle after three hundred and fifty millilitres have been used. How much was in the bottle originally? 625 ml
19 What is ten more than four hundred and ninety-seven? 507
20 Find the total of three ten-pence coins and five two-pence coins. 40p
21 Look at the numbers on your answer sheet. Draw a circle around each even number. 58, 50, 36
22 Look at the numbers on your answer sheet. Write them in order of size, starting with the largest. 866, 687, 551, 364
23 What is the total of six thousand, six hundred and sixty pounds, and three hundred and twenty pounds? £6980
24 Five people share forty-four video tapes equally. How many tapes are left over? 4
25 Add sixty-four and fifty-eight. 122

1 Write down the smallest number in this list: sixty-nine, forty, thirty-eight, fifty-seven, sixty-three. 38
2 What is one more than eighty-nine? 90
3 There are ten people living in a house. Six are children. How many are adults? 4
4 Write in figures: the number seven hundred and four. 704
5 What is the total value of three coins, if each coin has a value of two pence? 6p
6 Write in figures: the number three thousand and twenty-five. 3025
7 Thirty bones are to be shared equally between ten dogs in a kennel. How many bones will each dog receive? 3
8 Add together fourteen pounds, six pounds, and five pounds. £25
9 What is seven less than one hundred and three? 96
10 There are forty-seven houses in a road. Thirty-two receive post on one particular day. How many houses receive no post? 15
11 Mark has seventeen rabbits. He sells eleven of them. How many does he have left? 6
12 What is the greatest number of five-pence coins you could have for thirteen pence? 2
13 The temperature at dawn is three degrees. By noon, it has risen by thirteen degrees. What is the temperature at noon? 16°
14 How many minutes are there, between half past eleven and ten to twelve? 20 min
15 Look at the shape on your answer sheet. Draw lines to divide the shape into quarters.
16 Add twenty-nine and thirty-four. 63
17 What is ten more than four hundred and ninety-three? 503
18 Forty grams of spaghetti are shared between five people. How much does each receive? 8 g
19 What is seventeen less than fifty? 33
20 What is the time, twenty minutes before quarter past nine? 8.55*
21 Look at the numbers on your answer sheet. Draw a circle around each even number. 44, 52, 78
22 Look at the numbers on your answer sheet. Write them in order of size, starting with the largest. 986, 846, 561, 223
23 In the first post, eighty-three letters are delivered. There are twenty-eight letters in the second post. What is the total number of letters received that day? 111
24 Multiply two hundred and fifty-two by two. 504
25 From eighty-four, take away thirty-seven. 47

A31

Answers

1 Write down the largest number in this list: fifty-one, forty-eight, thirty-two, sixty-four, fifty-eight. **64**
2 What is one more than nine? **10**
3 Write in figures, the number six hundred and four. **604**
4 What are two twos? **4**
5 An artist has eight paintings to sell. He sells four. How many has he left? **4**
6 Write in figures: the number eight thousand, eight hundred and ninety. **8890**
7 There are eighteen nails in one box and twenty-one nails in another box. How many nails are there, altogether? **39**
8 Add together six, three, nine, and six. **24**
9 What is the number three hundred and fifty-one, to the nearest ten? **350**
10 Multiply six by three. **18**

11 There are four eggs in a bag. One quarter of the eggs are bad. How many good eggs are there? **3**
12 How many minutes are there, between five past five and twenty-five to six? **30 min**
13 Colin has eighteen litres of petrol at the start of a journey. He uses nine litres. How many litres remain? **9 litres**
14 A box weighs five kilograms. Its contents weigh twelve kilograms. What is the total weight of the box and its contents? **17 kg**
15 What is the greatest number of threepenny stamps you can buy for five pence? **1**
16 Find the total of three ten-pence coins and six two-pence coins. **42p**
17 Add thirty-three to forty-six. **79**
18 What is the time, fifteen minutes after ten to nine? **9.05***
19 What is twenty-one less than fifty-three? **32**
20 What is the total weight of five discs, each of weight twenty-five grams? **125 g**

21 Look at the numbers on your answer sheet. Write them in order of size, starting with the smallest. **85, 93, 143, 742**
22 Look at the numbers on your answer sheet. Draw a circle around each of the even numbers. **76, 66, 70**
23 An insurance account has two thousand, nine hundred and forty pounds in it. A bonus of two hundred pounds is also added. What is the total amount in the account? **£3140**
24 Add seventy-nine and thirty-two. **111**
25 Five people each pay thirty-five pounds for a day trip. What is the total amount paid by the five people? **£175**

A32

Answers

1 Write down the largest number in this list: fifty-eight, thirty-seven, twenty-four, forty-eight, thirty-two. **58**
2 What is two less than ninety-one? **89**
3 The are twelve used cars for sale. Five are sold in a week. How many are left? **7**
4 Write in figures: the number one hundred and one. **101**
5 Multiply two by four. **8**
6 There are thirteen boys and fourteen girls in a class. What is the total number of pupils in the class? **27**
7 Add together four, five, nine, and seven. **25**
8 What is the greatest number of two-pence coins you could have for sixteen pence? **8**
9 What are ten fours? **40**
10 Write in figures: the number eight thousand and seventy. **8070**

11 What is the greatest number of five-pence coins you could have for twenty-three pence? **4**
12 The temperature is sixteen degrees at noon, but is only seven degrees three hours later. By how many degrees has the temperature fallen? **9°**
13 Malika has nine zebra finches. She buys six more. How many zebra finches will she now have? **15**
14 What is the date, fourteen days after the fourth of March? **18th March**
15 Look at the line on your answer sheet. Draw an X halfway along the line. **(bisector)**
16 What is twelve less than forty-four? **32**
17 What is the time, ten minutes after five to eleven? **11.05***
18 Add thirty-seven and twenty-four. **61**
19 Three hundred grams of wheat grain are shared equally between five people. How much does each receive? **60 g**
20 There are thirty days in November. How many days are left after the seventeenth day? **13 days**

21 Look at the numbers on your answer sheet. Draw a circle around each of the odd numbers. **49, 75**
22 Look at the numbers on your answer sheet. Write them in order of size, starting with the smallest. **543, 856, 897, 898**
23 A salesman makes a journey of fifty-eight kilometres to the nearest city and fifty-eight kilometres back. What is the total distance travelled? **116 km**
24 Multiply three hundred and seventy-four by two. **748**
25 A book costs three pounds and seventy-five pence. Katy saves fifty pence each week. How many weeks will Katy have to save before she can buy the book? **8 wks**

1. Write down the smallest number in the list: sixty-three, eighty-six, twenty-seven, sixty-one, forty-four. — 27
2. What is two more than thirty-nine? — 41
3. Write in figures: the number one hundred and ten. — 110
4. What is eight, taken away from ten? — 2
5. Adam was born in nineteen ninety-two. How old was he in the year nineteen ninety-six? — 4 yrs
6. A prize of twenty pounds is shared equally between four people. How much does each one receive? — £5
7. Write the number twenty-eight, to the nearest ten. — 30
8. Eight pounds is knocked off the price of a television which would normally cost four hundred and four pounds. What is the reduced price? — £396
9. Write in figures: the number one thousand and seventy-two. — 1072
10. In a pool, there are twenty-three goldfish and six silver fish. How many fish are there, altogether? — 29

11. What is the greatest number of two-pence coins you can have for eleven pence? — 5
12. Neil draws a line fifteen centimetres long; he then increases the length of the line by another six centimetres. What is the final length of the line? — 21 cm
13. Philip buys eighteen eggs in a box. He drops the box and breaks five eggs. How many eggs does he have left? — 13
14. What is half of ten pence? — 5p
15. How many minutes are there, between quarter past six and twenty-five to seven? — 20 min
16. What is thirty-two, take away twenty-one? — 11
17. A mouse weighs one hundred and twenty-five grams. It has now doubled in weight. What is its new weight? — 250 g
18. What is eighteen less than thirty? — 12
19. How many seconds are there, in twenty minutes? — 1200
20. How much change do you get from ten pounds, if you spend six pounds and fifty pence? — £3.50

21. Look at the numbers on your answer sheet. Write them in order of size, starting with the smallest. — 53, 302, 813, 934
22. Look at the numbers on your answer sheet. Draw a circle around each odd number. — 19, 71
23. What is fifty-three, take away thirty-six? — 17
24. A coach makes five trips, each time with sixty-three passengers. How many passengers have been carried, altogether? — 315
25. Add seventy-five and forty-six. — 121

1. What is the largest number in this list: thirty-one, fifty-eight, forty-seven, twenty, forty-six? — 58
2. What is one less than twenty? — 19
3. There are five pairs of socks on a line. How many individual socks is this? — 10
4. Abdul has saved seven pounds. He spends three pounds. How many pounds has he left? — £4
5. Write in figures: the number four hundred and five. — 405
6. What is fifty, take away thirty-five? — 15
7. Two brothers are both aged thirteen years. What is their total age? — 26 yrs
8. What is the total of three kilograms, eight kilograms, and twelve kilograms? — 23 kg
9. What is nine shared by three? — 3
10. Write in figures: the number two thousand, six hundred and forty. — 2640

11. A school netball team scores six goals in their first game and eight goals in the next game. How many goals have been scored, altogether? — 14
12. A piece of wood is four centimetres long. One quarter is cut off. What length of wood has been cut off? — 1 cm
13. Kathryn takes twenty minutes to walk to the station and then waits eight minutes before the train arrives. How long is her journey from her house to the train? — 28 min
14. What is the greatest number of fourpenny sweets you could buy with nine pence? — 2
15. Jill has seven chairs in the lounge and nine in the dining room. How many chairs has she, altogether? — 16
16. What is ten more than four hundred and ninety-six? — 506
17. Find the total of three ten-pence coins and two five-pence coins. — 40p
18. Add forty-one and twenty-nine. — 70
19. A cake has been reduced from one pound and ninety-nine pence, to one pound and forty-nine pence. What is the actual reduction in its price? — 50p
20. Add twenty-nine and twenty-eight. — 57

21. Look at the numbers on your answer sheet. Write them in order of size, starting with the largest. — 982, 888, 745, 687
22. Look at the numbers on your answer sheet. Draw a circle around each odd number. — 71, 39, 67
23. Add thirty-seven and seventy-four. — 111
24. What is the cost of four video tapes costing two pounds and ninety-nine pence each? — £11.96
25. Add seven thousand, seven hundred and seventy grams, and four hundred grams. — 8170 g

1. Write down the smallest number in this list: sixty-seven, fifty-six, forty-two, seventy-two, fifty-three.   **42**
2. What is one more than forty-nine?   **50**
3. What are three threes?   **9**
4. Wai-Lee has nine stamps. She uses six of them to post some letters. How many stamps will she have left?   **3**
5. Write in figures: the number three hundred and sixty-five.   **365**
6. Add forty-three and twenty-two.   **65**
7. Write the number eighty-one, to the nearest ten.   **80**
8. There are thirty-four people on a bus. Twenty-seven people get off. How many are left on the bus?   **7**
9. Write in figures: the number one thousand and six.   **1006**
10. A man makes two journeys, each of length ten kilometres. What is the total distance travelled?   **20 km**

11. A café starts a day with twenty litres of orange juice. At the end of the day, eight litres remain. How many litres were used during the day?   **12 litres**
12. How many minutes are there, between ten past nine and twenty-five to ten?   **25 min**
13. How long was a holiday with seven days spent in one resort and eight days in a second resort?   **15 days**
14. Look at the shape on your answer sheet. Draw a line on the shape to divide it into halves.
15. What is the greatest number of two-pence coins you can have for eleven pence?   **5**
16. From thirty-nine, take twenty-five away.   **14**
17. There are fifty-four children in one swimming group and forty-six in another. How many children are learning to swim?   **100**
18. What is thirty, take away twelve?   **18**
19. How much remains of five pounds, if you spend three pounds and twenty-five pence?   **£1.75**
20. What is the time, fifteen minutes after ten to nine?   **9.05***

21. Look at the numbers on your answer sheet. Draw a circle around each even number.   **74, 72**
22. Look at the numbers on your answer sheet. Write them in order of size, starting with the smallest.   **577, 779, 977, 978**
23. There are fifty-three people in one train carriage and sixty-eight people in the next carriage. What is the total number of people in both carriages?   **121**
24. Multiply forty-seven by five.   **235**
25. Add four thousand, five hundred and fifty pounds, and six hundred pounds.   **£5150**

1. Write down the largest number in this list: sixty-three, forty-seven, fifty-three, thirty-two, forty-three.   **63**
2. Write in figures: the number seven hundred and twenty-three.   **723**
3. What is two less than twenty?   **18**
4. Sarah was born in nineteen ninety. How old was she in nineteen ninety-six?   **6 yrs**
5. There are six eggs in a box. Two are used to bake a cake. How many eggs are left?   **4**
6. How many hours are there, between half past seven in the morning and half past three in the afternoon?   **8 hrs**
7. What are ten tens?   **100**
8. Write in figures: the number four thousand and forty-five.   **4045**
9. A tuck shop has eighty packets of crisps. During the week, sixty-eight packets have been sold. How many are left?   **12**
10. Add thirty-seven and twenty-one.   **58**

11. The temperature is eight degrees now, but it was fifteen degrees earlier. By how many degrees has the temperature dropped?   **7°**
12. Remi has six litres of petrol in his tank. He buys another seventeen litres. How much petrol does he have now?   **23 litres**
13. How many minutes are there, between twenty to twelve and five to twelve?   **15 min**
14. There are ten sweets in a packet. Half the sweets are eaten. How many sweets are left?   **5**
15. What is the greatest number of threepenny sweets you could buy with ten pence?   **3**
16. Add forty-three and twenty-seven.   **70**
17. What is ten more than six hundred and ninety-two?   **702**
18. How many seconds are there, in five minutes?   **300 sec**
19. What is forty, take away twenty-four?   **16**
20. A coat costing forty-four pounds has its price halved. What is the new price?   **£22**

21. Look at the numbers on your answer sheet. Write them in order of size, starting with the smallest.   **789, 806, 832, 866**
22. Look at the numbers on your answer sheet. Draw a circle around each odd number.   **85, 51, 93**
23. There are fifty-three people on a coach. Thirty-nine get off. How many people remain?   **14**
24. Multiply two hundred and sixty-four by two.   **528**
25. Add sixty-seven and forty-four.   **111**

1 What is the largest number in this list: sixty-seven, fifty-five, fifty-two, twelve, eighteen?   67
2 What is two less than seventy?   68
3 There are two pens in a box. How many pens will there be in five similar boxes?   10
4 Write in figures: the number six hundred and five.   605
5 From ten, take five away.   5
6 What is the greatest number of five-pence coins you could have for eighteen pence?   3
7 A school is open forty weeks in a year. Twenty-eight weeks have passed. How many weeks remain?   12 wks
8 Write two hundred and seventy-six, to the nearest hundred.   300
9 What are three fives?   15
10 Write in figures: the number three thousand, six hundred and five.   3605

11 Ruth buys a pack of twenty-five envelopes. She uses sixteen of them. How many does she have left?   9
12 What is the greatest number of fourpenny stamps you can buy for eighteen pence?   4
13 Keith scores fifteen points in a game. Mark scores nine points. How many points did they score between them?   24
14 How many minutes are there, between ten past six and half past six?   20 min
15 Look at the line shown on your answer sheet. Divide this line into quarters.   (bisectors)
16 Add these numbers: eleven, twenty-two, thirteen.   46
17 There are one hundred and nine people at a disco. Ninety-one arrived at the start. How many were late?   18
18 What is nine more than three hundred and ninety-three?   402
19 A watch is five minutes slow. When the watch shows half past one, what is the correct time?   1.35*
20 What is thirteen less than thirty-five?   22

21 Look at the numbers on your answer sheet. Write them in order of size, starting with the largest.   365, 202, 109, 86
22 Look at the numbers on your answer sheet. Draw a circle around each even number.   44, 92, 64
23 Mrs Williams buys five litres of milk a day. How much milk will she buy in twenty-six days?   130 litres
24 Add seventy-five and fifty-five.   130
25 A tank contains one thousand, five hundred and fifteen litres of oil. Another seven hundred litres are added. How much oil is now in the tank?   2215 litres

1 Write down the smallest number in this list: forty, fifty-eight, forty-five, thirty-five, forty-four.   35
2 Write in figures: the number seven hundred and seventy-two.   772
3 What is one more than fifty-nine?   60
4 What are three twos?   6
5 Lisa has saved nine pounds. She spends four pounds. How much money has she left?   £5
6 A clock normally costing one hundred and three pounds has been reduced in price by five pounds. What is the reduced price?   £98
7 Write in figures: the number seven thousand and five.   7005
8 In a board game, twelve cards are to be shared equally between three players. How many cards does each player receive?   4
9 Add eighty-one and seventeen.   98
10 What is the greatest number of two-pence coins you could have for eighteen pence?   9

11 Harry cycles nine kilometres to the town, five kilometres to his sister's house and, then, eight kilometres home. How far does he cycle, altogether?   22 km
12 What is the date, sixteen days after the fourth of July?   20th July
13 A candle has a height of seventeen centimetres. It slowly burns down by nine centimetres. What will be the height of the candle now?   8 cm
14 What is half of four pounds?   £2
15 What is the greatest number of fourpenny sweets you could buy for fifteen pence?   3
16 What is twenty-three less than forty?   17
17 How much change do you get from ten pounds, if you spend eight pounds and sixty pence?   £1.40
18 What is ten more than four hundred and ninety-five?   505
19 Add together these numbers: eighteen, twenty, twelve.   50
20 A clock is five minutes slow. When the watch shows nine o'clock, what is the correct time?   9.05*

21 Look at the numbers on your answer sheet. Draw a circle around each odd number.   23, 15, 21
22 Look at the numbers on your answer sheet. Write them in order of size, starting with the largest.   977, 967, 683, 564
23 In a sale, a set of ladders normally costing forty-eight pounds is reduced by nineteen pounds. What is the sale price?   £29
24 Add forty-eight and seventy-three.   121
25 A toy costs two pounds and ninety pence. Jonathan saves fifty pence each week. How long will Jonathan have to save before he can buy the toy?   6 wks

1 Write down the smallest number in this list: thirty-seven, forty-two, thirty-five, twenty-two, twenty-nine.   **22**

2 Write in figures: the number six hundred and forty.   **640**

3 What is one less than eighty?   **79**

4 There are ten swings in a park. Seven are for older children. How many are for younger children?   **3**

5 You have two coins, each of which are five-pence coins. What is the total value of the two coins?   **10p**

6 There are ten pairs of shoes in a changing room. How many individual shoes is this?   **20**

7 How many hours are there, between one o'clock in the afternoon and seven o'clock in the evening?   **6 hrs**

8 Add fifty-three to thirty-four.   **87**

9 There are forty adults watching a play. Twenty-two are women. How many are men?   **18**

10 Write in figures: the number three thousand, one hundred and twenty.   **3120**

---

11 What is the greatest number of ten-pence coins you can have for twenty-five pence?   **2**

12 There are eighteen paintings in a gallery. Eleven are sold. How many are left?   **7**

13 In a class, there are eleven boys and thirteen girls. How many are there, in the class?   **24**

14 How many minutes are there, between twenty to three and three o'clock?   **20 min**

15 Look at the line on your answer sheet. Draw an X halfway along the line.   **(bisector)**

16 Add together these numbers: twenty, twenty-five, fourteen.   **59**

17 What is the time, twenty minutes after ten to eight?   **8.10***

18 What is thirty-one less than sixty-four?   **33**

19 A rabbit weighed one hundred and seventy-five grams. The rabbit has now doubled in weight. What is its new weight?   **350 g**

20 Add twenty-seven and thirty-six.   **63**

---

21 Look at the numbers on your answer sheet. Draw a circle around each even number.   **16, 24, 18**

22 Look at the numbers on your answer sheet. Write them in order of size, starting with the smallest.   **488, 897, 986, 987**

23 Derek has forty-eight five-pound notes. What is the total amount of money that Derek has?   **£240**

24 Add thirty-six and ninety-five.   **131**

25 Bill has eight thousand, eight hundred and forty pounds in his savings account. He adds a further three hundred to the account. How much money will there now be, in the account?   **£9140**

1 Write down the largest number in this list: seventy-four, fifty, fifty-two, sixty-eight, thirty-six.   **74**

2 What is one more than eighty-nine?   **90**

3 Write in figures: the number one hundred and sixty-nine.   **169**

4 A puppy was bought in nineteen ninety-two. How old was the puppy in nineteen ninety-seven?   **5 yrs**

5 What is five less than eight?   **3**

6 Write the number three hundred and nine, to the nearest hundred.   **300**

7 Add thirty-four to fourteen.   **48**

8 Write in figures: the number four thousand and four.   **4004**

9 Of twenty-three parents at a parents evening, nineteen had arrived by car. How many had not travelled by car?   **4**

10 There are five tennis balls in a carton. How many tennis balls will there be in four similar cartons?   **20**

---

11 There are eighteen cats in a cattery. Twelve are female. How many are male?   **6**

12 A pie has been cut into quarters. How many pieces are there?   **4**

13 What is the greatest number of five-pence coins you could have for nineteen pence?   **3**

14 What is the date, eleven days after the twentieth of May?   **31st May**

15 How many minutes are there, between twenty to six and six o'clock?   **20 min**

16 What is twenty-four less than sixty?   **36**

17 A radio costing eighty-eight pounds has its price halved. What is the new price?   **£44**

18 What is ten more than two hundred and ninety-three?   **303**

19 What is the time, fifteen minutes after five to four?   **4.10***

20 Add together these numbers: nineteen, twenty, eleven.   **50**

---

21 Look at the numbers on your answer sheet. Write them in order of size, starting with the smallest.   **500, 531, 578, 590**

22 Look at the numbers on your answer sheet. Draw a circle around each odd number.   **45, 41, 49**

23 What is sixty-two, take away twenty-nine?   **33**

24 Five people share twenty-seven crystal balls equally. How many crystal balls are left over?   **2**

25 Add eighty-seven and twenty-five.   **112**

B1

1 A dinner ticket cost three pounds. What is the total cost of five dinner tickets? — £15

2 Write in figures: two thousand, one hundred and four. — 2104

3 Subtract twenty-five from forty. — 15

4 Three leaves weighed twelve grams, eight grams, and eleven grams. What was the total weight? — 31 g

5 What is the greatest number of two-pence coins you could get for eighteen pence? — 9

6 What is thirty-two, multiplied by ten? — 320

7 How many sevenths are there, in two whole ones? — 14

8 What is two squared? — 4

9 How many hundreds are there, in eight hundred? — 8

10 A blank video tape costs six pounds. What is the total cost of buying seven blank video tapes? — £42

---

11 Tom has thirty-eight pounds. He spends seventeen pounds. How much has he left? — £21

12 Find the total of three ten-pence coins, and ten one-pence coins. — 40p

13 Add together twenty-four, thirty-two, and eleven. — 67

14 What is the total cost of five cartridge refills at eighteen pence each? — 90p

15 Two weekly shopping bills are forty-six pounds and thirty-seven pounds. What is the total amount spent? — £83

16 What is twice thirty-four? — 68

17 What is four point two, added to three point one? — 7.3

18 What is seventeen, multiplied by five? — 85

19 From one hundred and twelve, take away twenty-four. — 88

20 A coin is flipped many times. Fifty-four per cent of the time, it landed heads up. What percentage of the time, did it land tails up? — 46%

---

21 Find the total of four ten-pence coins, seven five-pence coins, and two two-pence coins. — 79p

22 Katy needs five pounds seventy, but she has saved only three pounds fifty. How much more does she need? — £2.20

23 How many sixteens are there, in six hundred and forty? — 40

24 What is thirty-five, multiplied by eleven? — 385

25 A train journey starts at eight thirty-five, and finishes at quarter past ten. How long did the journey take, in hours and minutes? — 1 hr 40 min

B2

1 Thirty counters are divided between ten people. How many will each person receive? — 3

2 A chewing gum costs four pence. How much will four chewing gums cost? — 16p

3 Write in figures: five thousand, one hundred and thirty. — 5130

4 What is sixty-three, added to thirteen? — 76

5 What is ninety-six to the nearest ten? — 100

6 What is five thousand, four hundred, divided by one hundred? — 54

7 Write down a multiple of three, which is less than ten. — 3, 6 or 9

8 What is the total cost of thirty-four premium bonds at one hundred pounds each? — £3400

9 What is forty-two days, written in weeks? — 6 wks

10 A large box of chocolates costs nine pounds. What would be the cost of eight boxes? — £72

---

11 Two pieces of wood have lengths of twenty-eight centimetres and thirty-three centimetres. What is their total length? — 61 cm

12 Add ten to two hundred and ninety-five. — 305

13 To twenty-one, add fourteen and, then, add eleven. — 46

14 Subtract fourteen from twenty-nine. — 15

15 Keith scores sixty-seven points playing a game but, then, loses twenty-nine points. How many points will he now have? — 38

16 Multiply forty-four by seven. — 308

17 How many halves are there, in six whole ones? — 12

18 A car journey starts at nine twenty, and lasts for fifty minutes. At what time, did the journey finish? — 10.10*

19 How many full boxes of six eggs each can be packed from fifty eggs? — 8

20 From four, take away three point two. — 0.8

---

21 John has four ten-pence coins and three five-pence coins. How much has John, altogether? — 55p

22 To seventy-eight, add fifty-three. — 131

23 How many thirteens are there, in three hundred and ninety? — 30

24 How much change do you get from twenty pounds, if you buy three shirts costing four pounds, fifty pence each? — £6.50

25 What is thirty-three, multiplied by thirteen? — 429

83

1. Tom and Mary share twelve pounds equally between them. How much does each receive? £6
2. What are ten fives? 50
3. Write in figures: one thousand and thirty-two. 1032
4. Subtract fifty-four from sixty. 6
5. From fourteen litres, take away nine litres and, then, add eleven litres. 16 litres
6. How many fifths are there, in four whole ones? 20
7. What is seven, multiplied by eight? 56
8. A box contains ten pens. How many pens are there, altogether, in sixty-seven boxes? 670
9. How many rows of nine chairs each can be set out using fifty-four chairs? 6
10. What is seventy-eight, divided by ten? 7.8

11. There are two hundred and thirty-five packets of biscuits on a supermarket shelf, and one hundred and four in the store room. How many packets are there, altogether? 339
12. From seventy, take away thirty-four. 36
13. Add together thirty-two, fourteen, and twenty-three. 69
14. Add ten to eight hundred and ninety-three. 903
15. Over two months, Jerry gained twenty-one points, and thirty-six points. How many is this, altogether? 57
16. A television programme starts at quarter past five, and finishes at ten past six. How long is the programme? 55 min
17. From one hundred and three, subtract thirty-six. 67
18. What are four lots of twenty-three? 92
19. To one point one, add one point nought one. 2.11
20. What is the greatest number of ten-pence coins you could get for fourteen pounds? 140

21. Five people share fifty-eight pound coins equally. How many pound coins are left over? 3 pound coins
22. To fifty-four, add forty-six. 100
23. Subtract three point nought two from four. 0.98
24. How many twenties are there, in three hundred? 15
25. What is twenty-two, multiplied by twelve? 264

84
Answers

1. Forty-five golf balls are shared equally between five players. How many golf balls does each player receive? 9
2. Write in figures: seven thousand and fifteen. 7015
3. Subtract nineteen from twenty-three. 4
4. Add forty-four to twenty-four. 68
5. What is three hundred and nine, to the nearest hundred? 300
6. What is six squared? 36
7. What is ninety-six, multiplied by one hundred? 9600
8. How many six-seater rowing boats will be needed for forty-two women? 7
9. Divide six hundred by ten. 60
10. What is one fifth of thirty? 6

11. Martin has thirty-four pounds. He spends eighteen pounds. How much has he left? £16
12. Add fifty-five to twenty-six. 81
13. One section of fencing measures ninety-five centimetres. What is the total length, in centimetres, of ten sections? 950 cm
14. Add together twenty-three, fourteen, and twenty-one. 58
15. There are one hundred and fifty-six Year 11 students at a school. One hundred and twenty-eight leave. How many students remain? 28
16. Multiply forty-one by eight. 328
17. A light in a room is switched on for sixty-eight per cent of a day. For what percentage is it switched off? 32%
18. What is one hundred and fifteen, subtract twenty-six? 89
19. What is two point three, added to nought point one four? 2.44
20. A bus journey starts at twenty to eight, and lasts for twenty minutes. At what time, does the journey finish? 8.00*

21. Jim has three ten-pence coins and four five-pence coins. How much has Jim, altogether? 50p
22. Add seventy-six and thirty-four. 110
23. How many seventeens are there, in five hundred and ten? 30
24. What is seventy-two, multiplied by eleven? 792
25. How much change should you get from twenty pounds, if you buy four books costing four pounds, fifty pence each? £2

1 How many hours are there, between one o'clock and eleven o'clock?   10 hrs

2 Write in figures: two thousand and four.   2004

3 What is twenty-five, divided by five?   5

4 Subtract thirty-two from fifty.   18

5 A family spent fifty-three pounds on Saturday, and thirty-four pounds on Sunday. How much did they spend, altogether, over the weekend?   £87

6 What is seven hundred and eighty-five, divided by ten?   78.5

7 What is two times two times two?   8

8 A box contains six eggs. How many eggs are there, altogether, in eight boxes?   48

9 Multiply thirty-six by one hundred.   3600

10 Write down a factor of nine.   1, 3 or 9

11 Mary has two ten-pence coins and two two-pence coins. How much has Mary, altogether?   24p

12 Subtract fifty-eight from seventy-two.   14

13 The number of pupils in three classrooms is eighteen, twenty-one, and thirty. What is the total number of pupils?   69

14 In a sale, the price of a hat is reduced from nine pounds ninety-nine to five pounds forty-nine. How much is the reduction?   £4.50

15 Add twenty-eight to forty-seven.   75

16 What is twice forty-two?   84

17 Subtract one point one from three point two.   2.1

18 How many thirds are there, in nine whole ones?   27

19 How many twenty-pence coins will add up to make eight pounds?   40

20 A box contains five sticks of gum. How many sticks of gum will there be, altogether, in thirty-six boxes?   180

21 What is the total length, in millimetres, of two sections of wire which have a length of six thousand, three hundred and ninety millimetres, and eight hundred millimetres?   7190 mm

22 Add thirty-seven to fifty-eight.   95

23 How many twelves are there, in two hundred and forty?   20

24 What is twenty-three, multiplied by fifteen?   345

25 A cinema programme lasting two hours and twenty-five minutes, starts at ten to three. At what time, does it finish?   5.15*

1 Three lengths of ten, nine, and four millimetres are put end to end. What is their total length?   23 mm

2 Add eighty-one to seventeen.   98

3 What is sixteen, divided by four?   4

4 Write in figures: one thousand, three hundred and seventy.   1370

5 Take five from one hundred and three.   98

6 A family needs nine pints of milk a day. How many pints will they need in a week?   63 pints

7 What is eighty-five, multiplied by ten?   850

8 What is seven squared?   49

9 What is four hundred and forty, divided by one hundred?   4.4

10 How many boxes of six eggs each can be packed from forty-eight eggs?   8

11 Ben has saved forty pounds. He spends fourteen pounds. How much has he left?   £26

12 Add together twenty-three, twenty, and thirty-five.   78

13 What is ten more than four hundred and ninety-eight?   508

14 Julie has sixty-four holiday photographs. Ann has twenty-five. How many holiday photographs have they, altogether?   89

15 What is the total cost of five packets of mints at twenty-nine pence each?   £1.45

16 What is twice forty-seven?   94

17 Add one point three to five point four.   6.7

18 How many quarters are there, in eight whole ones?   32

19 Jeremy starts his homework at half past seven, and finishes at quarter past eight. How long has he spent on his homework?   45 min

20 Multiply fifty-three by nine.   477

21 Find the total of three ten-pence coins and twelve two-pence coins.   54p

22 Ali puts two hundred pounds into an account which already has two thousand, eight hundred and four pounds in it. How much is there, altogether?   £3004

23 Subtract six point nought five from seven.   0.95

24 How many fifteens are there, in four hundred and fifty?   30

25 What is thirty-one, multiplied by twelve?   372

87

1 A glove has five fingers.
   How many fingers are there, altogether,
   on nine gloves?   45
2 Write in figures: two thousand and one.   2001
3 Jimmy and Tom share fourteen stamps
   equally. How many stamps does
   Jimmy receive?   7
4 What is the greatest number of
   five-pence coins you can get from
   eighteen pence?   3
5 Subtract twenty-eight from forty.   12
6 Cubes are packed in boxes of one
   hundred. How many boxes will be
   needed for eight thousand cubes?   80
7 Multiply eight by nine.   72
8 How many eighths are there, in three
   whole ones?   24
9 What is the total of thirty-nine
   ten-pound notes?   £390
10 What is nine squared?   81

11 Find the total of two ten-pence coins,
   two five-pence coins, and two
   two-pence coins.   34p
12 Add together twenty-one, twenty, and
   nineteen.   60
13 Paul has saved six pounds thirty-five
   pocket money, and is given another
   two pounds and forty pence. How much
   will he now have, altogether?   £8.75
14 Subtract thirty-four from seventy.   36
15 My watch is five minutes fast.
   When the watch shows half past ten,
   what is the correct time?   10.25*
16 What is a half, written as a percentage?   50%
17 Multiply sixty-two by three.   186
18 Seventy-three per cent of pupils in a
   room have blue eyes. What percentage
   do not have blue eyes?   27%
19 Subtract forty-three from one hundred
   and eight.   65
20 To three point one, add nought
   point two.   3.3

21 Find the total length, when seven
   hundred metres is added to five
   thousand, five hundred metres.   6200 m
22 What is twice thirty-six?   72
23 How many fourteens are there, in two
   hundred and eighty?   20
24 Multiply thirteen by thirteen.   169
25 An Irish ferry crossing lasts for two
   hours and fifty-five minutes. The ferry
   leaves Holyhead at half past seven.
   At what time, does it arrive in Dublin?   10.25*

88

1 On Saturday, there were forty-seven
   members at a youth club. On Sunday,
   there were thirty-one. What was the
   total attendance over the weekend?   78
2 Write in figures: three thousand
   and thirty.   3030
3 What is fifty-eight less than seventy?   12
4 How many hours are there, between
   oh seven thirty hours and fifteen thirty
   hours?   8 hrs
5 How many quarters are there, in two
   whole ones?   8
6 What is fifty-seven, multiplied by one
   hundred?   5700
7 To sixteen metres, add thirteen metres
   and, then, take nine metres away.   20 m
8 How many days are there, in six
   weeks?   42 days
9 What is four hundred and eighty,
   divided by ten?   48
10 What is three times three times three?   27

11 A hose is cut into three equal lengths
   of eight metres. What was the original
   length of the hose?   24 m
12 From sixty-four, take twenty-three.   41
13 Find the total of twenty-five, eleven,
   and twenty-one.   57
14 What is ten more than eight hundred
   and ninety-seven?   907
15 One brother has a collection of sixty-six
   football cards. The other has thirty-three.
   How many football cards have they,
   altogether?   99
16 What is twice sixty-three?   126
17 Subtract two point three from four
   point eight.   2.5
18 Multiply twenty-two by nine.   198
19 Thirty-eight per cent of people do most
   of their shopping at the weekend.
   What percentage of people do most of
   their shopping during the weekdays?   62%
20 A train journey starts at thirteen twenty
   hours, and last fifty minutes. At what
   time, does the journey finish?   14:10*

21 Add five hundred pounds to one
   thousand, eight hundred and
   fifty pounds.   £2350
22 How many thirteens are there, in five
   hundred and twenty?   40
23 Add twenty-nine to eighty-seven.   116
24 What is thirty-two, multiplied by eleven? 352
25 How much change should you get from
   thirty pounds, if you buy four tapes
   costing six pounds and twenty-five
   pence each?   £5

1	Twelve apples are shared equally between two people. How many does each receive?	6
2	Subtract twenty-seven from thirty-three.	6
3	Write in figures: seven thousand, seven hundred and seven.	7707
4	Sally earns twenty-one pounds on Monday, and forty-three pounds on Tuesday. How much does she earn, altogether?	£64
5	What is eighty-one, to the nearest ten?	80
6	What is three squared?	9
7	Write down a multiple of six, which is less than twenty.	6, 12 or 18
8	Multiply twenty-nine by one hundred.	2900
9	What is six hundred and seventy-four, divided by ten?	67.4
10	One half-dozen is six. How much is nine half-dozens?	54

11	Over a three-day period, the number of papers delivered by Rose is twenty-five, twenty-eight, and twenty-six. How many is this, altogether?	79
12	Subtract thirty-two from fifty.	18
13	What is twenty more than five hundred and eighty-three?	603
14	Add twenty-seven to seventy-three.	100
15	A shop reduced the cost of a packet of mackerel from one pound ninety-nine to one pound forty-nine. What is the reduction?	50p
16	From three point four two, subtract one point two one.	2.21
17	Write down the length of the perimeter of a square of side three centimetres.	12 cm
18	Multiply thirty-four by six.	204
19	Subtract forty-eight from one hundred and sixteen.	68
20	One sixth of a number is twelve. What is the number?	72

21	Find the total of five ten-pence coins, three five-pence coins, and one two-pence coin.	67p
22	Add nine hundred to one thousand, three hundred and twenty.	2220
23	From two, subtract one point nought five.	0.95
24	What is twelve, multiplied by fifteen?	180
25	How many eighteens are there, in three hundred and sixty?	20

1	John has forty pounds. He spends twenty-five pounds. How much has he left?	£15
2	Write in figures: three thousand, one hundred and four.	3104
3	How many tens are there, in seventy?	7
4	Add fourteen to fourteen.	28
5	An octagon has eight points. What is the total number of points for seven octagons?	56
6	Multiply eighty-two by ten.	820
7	How many fifths are there, in three whole ones?	15
8	Divide two hundred and forty by one hundred.	2.4
9	What is the total of three, seven, and twelve kilograms?	22 kg
10	What is five squared?	25

11	Two weights are fifty-six kilograms, and twenty-five kilograms. What is the total weight?	81 kg
12	Subtract twelve from thirty-eight.	26
13	What is ten more than seven hundred and ninety-one?	801
14	A clock is five minutes slow. When the clock is showing half past three, what in the correct time?	3.35*
15	What is the total cost of three tickets at four pounds and two tickets at five pounds?	£22
16	What is twice seventy-two?	144
17	Subtract one point five from six.	4.5
18	There are four counters in a game. How many counters are needed, altogether, for forty-two games?	168
19	What is the greatest number of ten-pence coins you could get for fifteen pounds?	150
20	Bill starts digging his garden at quarter past two, and finishes at five past three. For how long was Bill digging his garden?	50 mm

21	What is the year, three hundred years after the year seventeen fifty?	2050
22	What are three twenty-fives?	75
23	From three, subtract one point seven five.	1.25
24	How many fifteens are there, in seven hundred and fifty?	50
25	What is twenty-two, multiplied by thirteen?	286

1 Six cakes are divided equally between three children. How many does each child receive? **2**

2 Write in figures: two thousand and twenty-two. **2022**

3 Add sixteen to twenty-three. **39**

4 What is twenty-eight, to the nearest ten? **30**

5 What is eight less than four hundred and four? **396**

6 What is four hundred and forty-four, divided by one hundred? **4.44**

7 Write down a factor of six. **1, 2, 3 or 6**

8 A tray contains six tins of beans. How many tins of beans are there, altogether, on eight trays? **48**

9 What is twenty-five, multiplied by ten? **250**

10 How many nine-centimetre lengths can be cut from a 63-centimetre length of wood? **7**

11 Martin has driven forty-three miles in a day, and his wife has driven seventeen miles. What is their total joint mileage? **60 miles**

12 Subtract forty-five from ninety. **45**

13 The rainfall last year was sixty-three centimetres. This year, the rainfall is fourteen centimetres less. What is the rainfall this year? **49 cm**

14 Find the total of twenty-seven, twenty-one, and twenty. **68**

15 What is the time, twenty minutes after five to eleven? **11.15***

16 Multiply twelve by eight. **96**

17 Add two point nine to one point nought one. **3.91**

18 How many fifths are there, in ten whole ones? **50**

19 A man slept for sixteen per cent of a day. For what percentage was he awake? **84%**

20 Subtract twenty-seven from one hundred and six. **79**

21 What is the year, five hundred years before fourteen hundred and sixty? **960**

22 Multiply twenty-one by twelve. **252**

23 Add seventy-one to thirty-six. **107**

24 An aeroplane flight lasts from thirteen forty hours to sixteen twenty hours. How long was the flight, in hours and minutes? **2 hrs 40 min**

25 How many sixteens are there, in three hundred and twenty? **20**

1 What is the greatest number of two-pence coins you could get for fifteen pence? **7**

2 A car can seat four passengers. How many passengers can be carried in three similar cars? **12**

3 Write in figures: four thousand, one hundred and four. **4104**

4 Jenny has thirteen crayons, and Shamra has fourteen crayons. How many have they, altogether? **27**

5 Add together four, five, nine, and seven. **25**

6 Multiply forty-four by one hundred. **4400**

7 How many ninths are there, in two whole ones? **18**

8 Nine office workers have each won a prize of seven pounds. How much have they won, altogether? **£63**

9 Write down a multiple of seven, which is less than twenty. **7 or 14**

10 How many tens are there, in seven thousand? **700**

11 Paula has four ten-pence coins and five five-pence coins. How much has Paula, altogether? **65p**

12 Subtract thirty-one from forty-five. **14**

13 Adam has gained nineteen marks in Maths, and thirty-four in English. How many marks has he gained, altogether? **53**

14 Add ten to five hundred and ninety-six. **606**

15 How many seconds are there, in five minutes? **300 sec**

16 Multiply thirty-two by seven. **224**

17 What is three point four, add five point four? **8.8**

18 Louise has one hundred and fourteen pounds. She spends fifty-one pounds. How much has she left? **£63**

19 What is the greatest number of twenty-pence coins you could get for ten pounds? **50**

20 A lesson started at twenty to ten, and finished thirty-five minutes later. At what time, did it finish? **10.15***

21 What is the year, three hundred years before the year two thousand and ten? **1710**

22 What is twice fifty-six? **112**

23 From two, subtract one point seven five. **0.25**

24 What is fifty-two, multiplied by eleven? **572**

25 How much change do you get from twenty pounds, if you buy three ties at four pounds fifty pence each? **£6.50**

1. A school has just four lessons each day. How many lessons will it have over the five-day working week?   20
2. Write in figures: two thousand and ten.   2010
3. Subtract fifty-eight from seventy.   12
4. How many hours are there, between oh six fifteen hours and fourteen fifteen hours?   8 hrs
5. What is seven less than five hundred and five?   498
6. How many hundreds are there, in two thousand, five hundred?   25
7. What is eight squared?   64
8. There are fifty-five chocolate bars in a box. What is the total number of bars in ten chocolate boxes?   550
9. How many tenths are there, in three whole ones?   30
10. What is nine sixes?   54

11. Alison has fifty records. She sells twenty-two in a car boot sale. How many records has she left?   28
12. Add together twenty-five, twenty, and twenty-four.   69
13. From sixty-four, subtract thirty-seven.   27
14. Ali is exactly twice the height of his brother, who is seventy centimetres tall. How tall is Ali? Give your answer in centimetres.   140 cm
15. Add thirty-three to sixty-four.   97
16. What is sixty-seven, multiplied by three?   201
17. What is four, subtract three point one?   0.9
18. What is the greatest number of ten-pence coins you could get for eighteen pounds?   180
19. A plane began its descent at twenty thirty-five hours, and landed at twenty-one fifteen hours. How long did it take to descend and land?   40 min
20. Subtract thirty-nine from one hundred and eighteen.   79

21. Find the total of two ten-pence coins, three five-pence coins, and four two-pence coins.   43p
22. How many twelves are there, in three hundred and sixty?   30
23. Add forty-seven to fifty-nine.   106
24. How much change should you get from fifty pounds, if you buy three concert tickets at twelve pounds fifty pence each?   £12.50
25. Multiply twenty-one by fourteen.   294

1. Jumo has nine pairs of socks. How many individual socks is this, altogether?   18
2. Write in figures: nine hundred and one.   901
3. Twelve chocolates are shared equally between four children. How many does each child receive?   3
4. What is eighteen, add twenty-one?   39
5. What is three hundred and fifty-one, to the nearest hundred?   400
6. What is seventeen, multiplied by one hundred?   1700
7. Four out of every ten people on a bus are male. Write this as a fraction. $\frac{4}{10}$ or $\frac{2}{5}$
8. What is seven, multiplied by six?   42
9. What is four squared?   16
10. What is six hundred and seventy-three, divided by ten?   67.3

11. Ben has a collection of thirty-seven stamps, but sells nineteen of them. How many does he have left?   18
12. Add twenty-eight to fifty-one.   79
13. What is the time, ten minutes before five past two?   1.55*
14. On Monday, sixty-eight pupils were late. On Tuesday, twenty-two were late. How many pupils were late, altogether?   90
15. What is ten more than nine hundred and ninety-six?   1006
16. What is twice eighty-six?   172
17. What is four point eight, subtract two point seven?   2.1
18. One seventh of a number is nine. What is the number?   63
19. There are seven days in a week. How many days are there, in forty-five weeks?   315 days
20. Out of a hundred people, nineteen travelled regularly on a train. Write this as a percentage.   19%

21. Find the total of three ten-pence coins, three five-pence coins, and three two-pence coins.   51p
22. What year is four hundred years before the year thirteen thirty?   930
23. From six, subtract two point seven five.   3.25
24. Multiply twenty-two by thirteen.   286
25. How many seventeens are there, in three hundred and forty?   20

B15

1. There are eight packets of cereal in a box. How many packets of cereal are there, in five boxes? — 40
2. Write in figures: one thousand and fourteen. — 1014
3. Subtract thirty-one from forty-seven. — 16
4. Add together fourteen pounds, seven pounds, and six pounds. — £27
5. From one hundred and three, subtract five. — 98
6. What is nine squared? — 81
7. A coach can carry sixty-three passengers. What is the greatest number of passengers that ten similar coaches can carry? — 630
8. What is eight sevens? — 56
9. How many sevenths are there, in three whole ones? — 21
10. What is one thousand, seven hundred and fifty, divided by one hundred? — 17.5

11. Alan has sixty-nine football cards. He buys another twelve. How many will he now have? — 81
12. Add twenty to six hundred and eighty-five. — 705
13. Subtract twenty-nine from thirty-seven. — 8
14. A watch is ten minutes fast. When the watch shows twenty to six, what is the correct time? — 5.30*
15. What is twenty-four, multiplied by four? — 96
16. What is five point nought one, added to three point nought four? — 8.05
17. How much change do you get from five pounds, if you spend four pounds thirty-five? — £0.65
18. Write down the length of the perimeter of a square of side four centimetres. — 16 cm
19. Subtract fifty-five from one hundred and four. — 49
20. A video machine taped a forty-five minute programme, starting at twenty thirty hours. At what time, did it finish taping? — 21.15*

21. Find the total of four ten-pence coins, five five-pence coins, and six two-pence coins. — 77p
22. What is twenty-five, multiplied by three? — 75
23. Add twenty-one to eighty. — 101
24. How many fifteens are there, in six hundred? — 40
25. Multiply twelve by eleven. — 132

B16

1. A car ferry takes twenty-seven cars one way and, then, brings seventy-one back the other way. How many cars is this, altogether? — 98
2. Write in figures: five thousand and forty. — 5040
3. From fifty, subtract thirty-six. — 14
4. There are four pairs of gloves on a washing line. How many individual gloves is this, altogether? — 8
5. Add two, seven, six, and eight. — 23
6. Write down a multiple of two, which is less than ten. — 2, 4, 6 or 8
7. In an arena, there are one hundred rows, each with eighty-eight chairs. How many people can sit in the arena? — 880
8. Write down the answer to eight, multiplied by nine. — 72
9. How many quarters are there, in five whole ones? — 20
10. What is five hundred and forty, divided by ten? — 54

11. What is the total weight of five discs, each of which weighs twenty-five grams? — 125 g
12. Find the total of three ten-pence coins, and six two-pence coins. — 42p
13. Subtract twenty-one from sixty-three. — 42
14. Add twenty-six to thirty-three. — 59
15. What is the time, fifteen minutes after ten to nine? — 9.05*
16. From eight, subtract two point five. — 5.5
17. What is thirty-three, multiplied by eight? — 264
18. A box contains eight pens. How many full boxes can be packed from a supply of sixty pens? — 7
19. What is one point two, added to one point two five? — 2.45
20. A table tennis match started at five to four, and finished at twenty to five. How long was the game? — 45 min

21. A concert lasts from ten to nine, to quarter past eleven. How long is the concert, in hours and minutes? — 2 hrs 25 min
22. What is twice thirty-eight? — 76
23. How many elevens are there, in three hundred and thirty? — 30
24. Multiply thirty-one by twelve. — 372
25. You buy four books at five pounds and fifty pence each. What change should you receive from thirty pounds? — £8

1 A nursery school has eight twins. How many individual children is this?   16
2 Write in figures: three thousand and two.   3002
3 Twenty bottles of milk are shared equally between five people. How many bottles does each person receive?   4
4 Add together fifteen kilograms and nine kilograms and, then, take away seven kilograms.   17kg
5 What is fifteen, multiplied by one hundred?   1500
6 How many hours are there, between eight o'clock in the morning and noon?   4 hrs
7 What is two squared?   4
8 What is seven hundred and thirty-seven, divided by ten?   73.7
9 Write down a factor of fifteen.   1, 3, 5 or 15
10 What is nine, multiplied by six?   54

11 There are thirty people on a bus. Eighteen get off. How many people are left?   12
12 What is ten more than six hundred and ninety-four?   704
13 Add together twenty, twenty-three, and twenty-five.   68
14 There are sixty-eight apples in a box. Another thirty-two apples are put in the box. What is the total number of apples in the box?   100
15 What is twenty-six, multiplied by three?   78
16 How many seconds are there, in ten minutes?   600 sec
17 What is three point four, add four point three?   7.7
18 How many sixths are there, in eight whole ones?   48
19 On a certain day, nine per cent of pupils were absent from a school. What percentage were present?   91%
20 Subtract forty-seven from one hundred and ten.   63

21 Find the total of five ten-pence coins, five five-pence coins, and five two-pence coins.   85p
22 From eight, subtract six point two five.   1.75
23 To sixty-four, add twenty-eight.   92
24 Multiply twenty-two by thirteen.   286
25 How many twelves are there, in six hundred?   50

1 A charity event has built forty-one towers, each of one hundred coins. How many coins is this, altogether?   4100
2 From thirty, subtract eighteen.   12
3 Write in figures: one thousand, four hundred and ninety.   1490
4 Write the answer to four times eight.   32
5 To twenty-five, add fifty-two.   77
6 What is eight squared?   64
7 How many days are there, in six weeks?   42 days
8 Find the total of fifteen two-pence coins.   30p
9 Write down a multiple of four, which is less than fifteen.   4, 8 or 12
10 How many tens are there, in five thousand, one hundred?   510

11 A clock is twenty minutes slow. When the clock shows half past four, what is the correct time?   4.50*
12 From sixty-eight, subtract thirty-eight.   30
13 Add fifty-eight to thirty-five.   93
14 Add twenty, twenty-three, and twenty-six.   69
15 A box weighs six kilograms, and its contents weigh thirty-five kilograms. What is the total weight?   41 kg
16 What is four point five, subtract one point three?   3.2
17 Multiply fifty-seven by six.   342
18 One sixth of a number is nine. What is the number?   54
19 From one hundred and five, subtract thirty-eight.   67
20 Bill has a bike ride for forty minutes each day. If he left home at twenty-five minutes past eleven, at what time should he return?   12.05*

21 You buy three CDs costing ten pounds and forty pence each. What change should you receive from forty pounds?   £8.80
22 What is twice twenty-eight?   56
23 Subtract twenty-eight from fifty-four.   26
24 Divide six hundred and eighty by forty.   17
25 Multiply fifteen by fifteen.   225

B19

1 A pack of fifteen biscuits is shared equally between five people. How many biscuits does each person receive? 3
2 Write two thousand and forty-five in figures. 2045
3 What is three, multiplied by three? 9
4 There are twenty-two people in a tram car. Sixteen people get on. How many are now in the tram car? 38
5 What is twenty-six, to the nearest ten? 30
6 What is two hundred and forty-three, divided by one hundred? 2.43
7 Write down a factor of fourteen. 1, 2, 7 or 14
8 Write down the answer to seven, multiplied by eight. 56
9 What is seventy-nine, multiplied by ten? 790
10 How many tenths are there, in four whole ones? 40

11 Forty grams of spaghetti are shared between five people. How much does each receive? 8 g
12 Subtract seventeen from fifty. 33
13 What is ten more than three hundred and ninety-three? 403
14 What is the time, twenty minutes before quarter past nine? 8.55*
15 Add nineteen and thirty-three. 52
16 Multiply forty-three by nine. 387
17 To nought point three one, add nought point nought five. 0.36
18 Janice has one hundred and twelve pounds. She spends thirty-two pounds. How much has she left? £80
19 What is the greatest number of twenty-pence coins you could get for nine pounds? 45
20 A radio programme started at eighteen twenty hours, and finished at nineteen ten hours. How long was the programme? 50 min

21 Add together three thousand, eight hundred and ten pounds, and five hundred pounds. £4310
22 From seven, subtract four point two five. 2.75
23 Add seventy-nine and thirty-two. 111
24 How many fourteens are there, in seven hundred? 50
25 What is twenty-one, multiplied by fourteen? 294

B20

1 What is the greatest number of five-pence coins you could get for fifty pence? 10
2 Write in figures: three thousand and three. 3003
3 How many twos are there, in eight? 4
4 The temperature falls from twenty-seven degrees to nineteen degrees. By how many degrees has the temperature fallen? 8°
5 Add together five, nine, seven, and nine. 30
6 Divide five hundred and ninety by one hundred. 5.9
7 What is six squared? 36
8 How many days are there, in nine weeks? 63 days
9 How many minutes are there, in four hours? 240 min
10 Multiply fifty-two by one hundred. 5200

11 There are thirty-eight passengers in one coach, and fifty-two passengers in a second coach. How many passengers are there, altogether? 90
12 From forty-eight, take sixteen. 32
13 Find the total of three ten-pence coins, and five two-pence coins. 40p
14 What is ten more than seven hundred and ninety-seven? 807
15 There are two hundred and seventy-five millilitres left in a bottle, after three hundred and fifty millilitres have been used. How much was in the bottle originally? 625 ml
16 What is twice fifty-four? 108
17 From two, subtract nought point four. 1.6
18 A taxi can carry six people. How many taxis are needed for forty people? 7
19 What is fifty-five, multiplied by five? 275
20 Fifty-eight per cent of people in a week passed their driving test at a centre. What percentage failed? 42%

21 A car journey is expected to take three hours and ten minutes. If the journey starts at twenty past eleven, at what time should it finish? 2.30*
22 From nine, subtract seven point two five. 1.75
23 How many nineteens are there, in five hundred and seventy? 30
24 From forty-seven, subtract twenty-nine. 18
25 Multiply forty-two by twelve. 504

1 There are seven pairs of bookends on a shelf. How many individual bookends is this, altogether? 14
2 How many tens are there, in forty? 4
3 Write in figures: seven thousand and twenty. 7020
4 Add seventy-five to fifty-three. 128
5 What is six less than one hundred and two? 96
6 A packet contains eight pens. How many pens will there be in six packets? 48
7 Multiply twenty-three by ten. 230
8 Write down a multiple of ten, which is less than thirty. 10 or 20
9 What is four squared? 16
10 What is four hundred and twenty-seven, divided by one hundred? 4.27

11 There are thirty-two pupils in a class. Twenty-one leave for another lesson. How many pupils remain? 11
12 Add together twenty-three, twenty-four, and twenty-five. 72
13 How many seconds are there, in twenty minutes? 1200 sec
14 What is forty, subtract eighteen? 22
15 How much change should you get from ten pounds, if you spend six pounds and fifty pence? £3.50
16 What is twice eighty-three? 166
17 From five point four two, subtract three point four. 2.02
18 A bus journey started at five to seven, and lasted thirty-five minutes. At what time, did the journey finish? 7.30*
19 Work out twenty-five, multiplied by eight. 200
20 How many halves are there, in eight whole ones? 16

21 Find the total of five ten-pence coins, four five-pence coins, and three two-pence coins. 76p
22 Add fifty-eight to fifty-eight. 116
23 How many elevens are there, in two hundred and twenty? 20
24 Multiply twenty-one by fifteen. 315
25 What is the year, which is five hundred years after the year sixteen hundred and forty? 2140

1 Sixty people are in a train carriage. Fifty-two get off. How many people remain? 8
2 Write in figures: one thousand and ten. 1010
3 A tripod has three legs. How many legs in total are there, on two tripods? 6
4 Find the total distance of twelve kilometres, seven kilometres, and nine kilometres. 28 km
5 How many hours are there, between on six thirty hours and twenty thirty hours? 14 hrs
6 What is four hundred and twenty-three, divided by ten? 42.3
7 How many eighths are there, in four whole ones? 32
8 What are seven nines? 63
9 How many pence are there, in sixty pounds? 6000p
10 Three out of every four people in a class had blue eyes. Write this as a fraction. $\frac{3}{4}$

11 Three hundred grams of wheat corn are shared equally between five people. How much does each one receive? 60 g
12 What is twelve less than sixty-four? 52
13 Add thirty-seven and twenty-four. 61
14 What is forty, subtract seventeen? 23
15 What is the time, ten minutes after five to eleven? 11.05*
16 Multiply sixty-four by three. 192
17 To two point nought three, add one point three. 3.33
18 A driving test centre can arrange tests for nine people per hour. How many whole hours will it take to test sixty people? 7 hrs
19 What is forty-five less than one hundred and nine? 64
20 A lesson lasted from twenty-five minutes past nine to ten o'clock. How long was the lesson? 35 min

21 You buy four train tickets costing five pounds and fifty pence each. What change should you receive from thirty pounds? £8
22 What is twice thirty-seven? 74
23 Multiply twenty-two by thirteen. 286
24 From fifty-three, subtract twenty-seven. 26
25 How many fourteens are there, in four hundred and twenty? 30

B23
Answers

1 What is the greatest number of two-pence coins you could get for twenty-five pence? — 🕐 12
2 How many days are there, in five weeks? — 35 days
3 Write in figures: two thousand and thirty-three. — 2033
4 Two sisters are aged fourteen and sixteen. What is their combined age, in years? — 30 yrs
5 What is four hundred and thirty-nine, to the nearest hundred? — 400
6 What is three squared? — 9
7 How many hundreds are there, in one thousand, six hundred? — 16
8 Multiply seventy-four by ten. — 740
9 How many sixths are there, in five whole ones? — 30
10 There are nine dots on a card. How many dots in total are there, on eight similar cards? — 72

11 A cake has been reduced, from one pound ninety-nine pence to one pound forty-nine pence. What is the reduction? — 🕐 50p
12 Add twenty-nine to seventy-one. — 100
13 Find the total of three ten-pence coins, and two five-pence coins. — 40p
14 What is twenty-nine, increased by twenty-eight? — 57
15 What is ten more than six hundred and ninety-six? — 706
16 From five point four two, subtract three point nought two. — 2.4
17 What is a half, written as a percentage? — 50%
18 There was a power cut, from twenty past six to five past seven. How many minutes was this? — 45 min
19 What is thirty-five, multiplied by nine? — 315
20 What is four point four, added to four point nought four? — 8.44

21 A CD unit costs eighty-six pounds, and the speakers cost thirty-four pounds. What is the total cost? — 🕐 £120
22 How many elevens are there, in four hundred and forty? — 40
23 What is twice forty-six? — 92
24 From four, subtract one point seven five. — 2.25
25 Multiply twenty-one by fourteen. — 294

B24
Answers

1 Helen has twenty-three pounds. She spends seventeen pounds. How much has she left? — 🕐 £6
2 Write in figures: seven thousand, eight hundred and five. — 7805
3 How many fives are there, in thirty-five? — 7
4 What are six twos? — 12
5 What is five less than four hundred and four? — 399
6 How many quarters are there, in four whole ones? — 16
7 There are eight cartons of drink on a tray. How many cartons are there, on six trays? — 48
8 How many tens are there, in nine hundred and seventy? — 97
9 Multiply twenty-two by one hundred. — 2200
10 Write down a multiple of eight, which is less than twenty. — 8 or 16

11 There are one hundred and nine people at a dinner, but only ninety-one stay on for the disco. How many people leave? — 🕐 18
12 What is thirteen less than thirty-five? — 22
13 Add together twenty-one, twenty-two, and twenty-three. — 66
14 What is nine more than four hundred and ninety-three? — 502
15 A watch is five minutes slow. When the watch shows half past one, what is the correct time? — 1.35*
16 From three, subtract one point seven. — 1.3
17 Write down the length of the perimeter of a square of side five centimetres. — 20 cm
18 In a school, forty-seven per cent of pupils had a school dinner. What percentage did not have a school dinner? — 53%
19 Multiply twenty-one by seven. — 147
20 From one hundred and twenty, subtract thirty-five. — 85

21 A temperature of fifty-three degrees falls to thirty-seven degrees. By how many degrees has it fallen? — 🕐 16°
22 How many twelves are there, in seven hundred and twenty? — 60
23 Brian's first lesson in school is at ten past nine. His last lesson finishes at twenty-five minutes to four. For how long is he in school? Give your answer in hours and minutes. — 6 hrs 25 min
24 Find the total of four ten-pence coins, four five-pence coins, and four two-pence coins. — 68p
25 Multiply twenty-two by fifteen. — 330

33

B25

1   What is the total weight of seventeen grams, eight grams, and seven grams? 32 g

2   Write in figures: four thousand and seven. 4007

3   What are three twos? 6

4   I have been waiting for fifty-three minutes for an hour to pass. How long have I yet to wait? 7 min

5   What is seven less than eight hundred and two? 795

6   Write down a factor of fifteen. 1, 3, 5 or 15

7   How many pence are there, in ninety-four pounds? 9400p

8   There are nine dots on a card. How many dots will there be on six similar cards? 54

9   What is ten squared? 100

10 Divide two hundred and forty-five by ten. 24.5

11 The weight of two pens is fifty-four grams and forty-six grams. What is their total weight? 100 g

12 What is the time, fifteen minutes after eight fifty? 9.05*

13 Subtract twenty-five from forty-nine. 24

14 How much change should you get from five pounds, if you spend three pounds, twenty-five pence? £1.75

15 What is twelve less than forty? 28

16 What is twice fifty-two? 104

17 Of a hundred people questioned, thirty-six shopped at a local supermarket. What percentage is this? 36%

18 From four point four, subtract two point three. 2.1

19 How many thirds are there, in nine whole ones? 27

20 A chicken lays three eggs each hour. How many will she have laid after thirty-eight hours? 114

21 Lucy has spent four and a quarter hours at a theme park. She arrived at ten to eleven. At what time, does she leave? 15.05*

22 What is twice sixty-six? 132

23 What is thirty-six, subtracted from eighty-three? 47

24 How many elevens are there, in seven hundred and seventy? 70

25 Multiply thirty-four by twelve. 408

B26

1   A large load can be moved nine miles every hour. How far will it have been moved after eight hours? 72 miles

2   Write in figures: one thousand and thirty-four. 1034

3   How many threes are there, in nine? 3

4   Add fifty-one to twenty-three. 74

5   What is forty-four, to the nearest ten? 40

6   What is twenty-seven less than forty? 13

7   What is five squared? 25

8   A bus can carry forty-three passengers. How many passengers are there, in ten full buses? 430

9   How many ninths are there, in three whole ones? 27

10 What is three hundred and forty-three, divided by one hundred? 3.43

11 A picture costs forty-three pounds, and its frame costs twenty-seven pounds. What is the total cost? £70

12 What is ten more than eight hundred and ninety-two? 902

13 Subtract twenty-four from fifty. 26

14 How many seconds are there, in five minutes? 300 sec

15 A coat costing forty-four pounds has its price halved. What is the new price? £22

16 To one point one two, add three point nought four. 4.16

17 What is fifty-one, multiplied by six? 306

18 A carriage can seat six people. How many carriages are needed for a group of twenty-five people? 5

19 From one hundred and twenty-two, subtract eighty-five. 37

20 A football match was abandoned after fifty-five minutes due to bad weather. It started at twenty past seven. At what time, did it finish? 8.15*

21 In one year, a town had two thousand, eight hundred and thirty millimetres of rain. Seven hundred millimetres fell during the following six months. What was the total rainfall, in millimetres? 3530 mm

22 What is twenty-six less than eighty-three? 57

23 Add sixty-one to forty-seven. 108

24 How many sixteens are there, in four hundred and eighty? 30

25 What is twenty-one, multiplied by fourteen? 294

1 Martin walked eighteen kilometres in the morning and, then, twelve kilometres in the afternoon. What is the total distance he walked that day? **30 km**
2 Subtract nineteen from forty. **21**
3 How many tens are there, in fifty? **5**
4 Write in figures: two thousand, two hundred and four. **2204**
5 Add together eight, six, nine, and seven. **30**
6 What is nine squared? **81**
7 A crate contains ten milk bottles. How many milk bottles are there, in eighty-three crates? **830**
8 What are seven sixes? **42**
9 Write down a multiple of nine, which is less than thirty. **9, 18 or 27**
10 How many hundreds are there, in five thousand, nine hundred? **59**

11 A rabbit weighed one hundred and seventy-five grams. It has now doubled in weight. What is the new weight? **350 g**
12 Add together twenty, twenty-five, and twenty-four. **69**
13 What is thirty-one less than seventy-four? **43**
14 Add thirty-seven and thirty-six. **73**
15 What is the time, twenty minutes after ten to eight? **8.10***
16 Multiply forty-six by four. **184**
17 One fifth of a number is nine. What is the number? **45**
18 From four point four two, subtract one point nought two. **3.4**
19 Subtract fifty-seven from one hundred and twenty-four. **67**
20 In one particular year, thirty-four per cent of the days were wet days. What percentage were dry days? **66%**

21 Find the total of two ten-pence coins, four five-pence coins, and six two-pence coins. **52p**
22 How many elevens are there, in five hundred and fifty? **50**
23 Multiply fifteen by twelve. **180**
24 A power cut lasts from five past eleven until five to three. How long, in hours and minutes, was the power cut? **3 hrs 50 min**
25 Add eighty-two and fifty-six. **138**

1 How many hours are there, between thirteen forty hours, and twenty-three forty hours? **10 hrs**
2 Write in figures: two thousand, one hundred. **2100**
3 What are nine tens? **90**
4 Subtract thirty-seven from forty. **3**
5 A train ticket costs five pounds. What would be the cost of eight tickets? **£40**
6 Multiply eighty-seven by one hundred. **8700**
7 Six out of every ten people on a bus are making a short journey. Write this as a fraction. **$\frac{6}{10}$ or $\frac{3}{5}$**
8 How many tens are there, in three thousand, two hundred? **320**
9 What is eight, multiplied by seven? **56**
10 What is seven squared? **49**

11 A clock is running five minutes slow. When the watch shows nine o'clock, what is the correct time? **9.05***
12 Add eighteen, twenty, and twenty-two. **60**
13 What is twenty-three less than thirty? **7**
14 What is ten more than five hundred and ninety-five? **605**
15 How much change should you receive from ten pounds, if you spend eight pounds and forty pence? **£1.60**
16 To two point one three, add four point two three. **6.36**
17 How many halves are there, in three and a half hours? **7**
18 From one hundred and one, subtract fifty-three. **48**
19 Write down the perimeter of a square of side one centimetre. **4 cm**
20 What is sixty-five, multiplied by seven. **455**

21 Forty-four sheep are loaded onto a lorry. Twenty-seven are taken off. How many remain? **17**
22 What is twice thirty-seven? **74**
23 How many thirties are there, in five hundred and forty? **18**
24 A gardener takes four hours and twenty minutes to put up a fence. He finished the job at two o'clock. At what time, did he start? **9.40***
25 Multiply fifteen by eleven. **165**

1 There are six socks on a washing line. How many pairs of socks is this? — 3
2 Write in figures: three thousand and seven. — 3007
3 How many tens are there, in sixty? — 6
4 One chocolate bar costs twenty-six pence. Another costs twelve pence. What is the total cost? — 38p
5 Add four, six, nine, and eight. — 27
6 How many tenths are there, in two whole ones? — 20
7 Divide four hundred and thirty by one hundred. — 4.3
8 Write down a multiple of four, which is less than twenty. — 4, 8, 12 or 16
9 What are nineteen tens? — 190
10 Multiply nine by eight. — 72

11 One crate weighs twenty-eight kilograms. Another weighs fourteen kilograms. What is the total weight? — 42 kg
12 Find the total of four ten-pence coins and two five-pence coins. — 50p
13 What is twelve less than thirty-two? — 20
14 Add thirty to seven hundred and seventy-four. — 804
15 At a hockey match, there are five hundred and fifty people standing, and one hundred and thirty-five people sitting. What is the total number of people at the match? — 685
16 Multiply sixteen by nine. — 144
17 Tent pegs are sold in packets of eight. Sally needs sixty pegs. How many packets will she need to buy? — 8
18 From one hundred and nineteen, subtract fifty-three. — 66
19 To one point four, add two point five one. — 3.91
20 A school lunch break runs from twelve forty hours to thirteen thirty hours. How long is the lunch break? — 50 min

21 One jar contains three thousand, six hundred and forty millilitres of juice. Another contains six hundred millilitres of juice. What is the total volume of juice? — 4240 ml
22 From fifty-two, subtract thirty-seven. — 15
23 Add sixty-two to forty-five. — 107
24 How many fifteens are there, in nine hundred? — 60
25 Multiply forty-three by eleven. — 473

1 Sixteen chocolate buttons are to be shared equally between two children. How many buttons should each receive? — 8
2 Write in figures: one thousand and twenty-five. — 1025
3 Subtract thirty-seven from fifty. — 13
4 Three pieces of wood, of widths eighteen millimetres, seven millimetres, and five millimetres, are put on top of each other. What is the total width? — 30 mm
5 Divide two hundred and twenty-four by ten. — 22.4
6 It takes James ten hours to put a model kit together. How long will it take him to put six of these kits together? — 60 hrs
7 Multiply sixty-four by one hundred. — 6400
8 What is four squared? — 16
9 How many eighths are there, in two whole ones? — 16
10 There are eight thimbles in a box. How many thimbles are there, in six boxes? — 48

11 A radio costing eighty-eight pounds has its price halved. What is the new price? — £44
12 Add together eighteen, twenty, and twenty-two. — 60
13 What is the time, fifteen minutes after five to four? — 4.10*
14 Subtract twenty-four from seventy. — 46
15 What is ten more than two hundred and ninety-three? — 303
16 From eight, subtract six point two. — 1.8
17 Multiply twenty-seven by five. — 135
18 One hundred pupils went on a school trip. Forty-four brought their own lunch. Write this as a percentage. — 44%
19 Subtract thirty-three from one hundred and two. — 69
20 A train journey starts at quarter past six, and lasts for forty-five minutes. At what time, does the journey finish? — 7.00*

21 Sixty-four people are on a plane. Thirty-six get off. How many remain? — 28
22 What is twice thirty-seven? — 74
23 How many elevens are there, in eight hundred and eighty? — 80
24 Multiply twenty-three by thirteen. — 299
25 Find the total of five ten-pence coins, two five-pence coins, and two two-pence coins. — 64p

B31　Answers

1　There are ten beads on each row of an abacus. There are eight rows.
　How many beads are there, altogether?　80
2　How many fives are there, in forty?　8
3　Write in figures: two thousand, two hundred.　2200
4　What is five hundred and sixty-two, to the nearest hundred?　600
5　There are seventy-three people at a play and, then, another sixteen arrive.
　What is the total attendance?　89
6　How many hundreds are there, in four thousand, three hundred?　43
7　What is seven squared?　49
8　Write down a factor of four.　1, 2 or 4
9　Multiply fifty-three by one hundred.　5300
10　How many days are there, in eight weeks?　56

11　A pair of trousers costs twenty-seven pounds, and a matching shirt costs fourteen pounds.
　How much does the outfit cost, altogether?　£41
12　What is ten more than six hundred and ninety-eight?　708
13　Add forty-five and fifty-five.　100
14　One bottle of shampoo contains one hundred and twenty-five millilitres.
　How much will two bottles contain?　250 ml
15　How many seconds are there, in ten minutes?　600 sec
16　What is twice thirty-three?　66
17　From five point nought four, subtract one point nought two.　4.02
18　How many fifths are there, in five whole ones?　25
19　Multiply thirty-one by eight.　248
20　Seventy-one per cent of the spectators at a football match were male.
　What percentage were female?　29%

21　An Intercity train leaves Glasgow at eleven forty hours, and arrives at London Euston at sixteen fifty hours.
　How long, in hours and minutes, is the journey?　5 hrs 10 min
22　Add eighty-three and thirty-four.　117
23　From eight, subtract six point two five.　1.75
24　Multiply thirty-four by twelve.　408
25　How many thirteens are there, in two hundred and sixty?　20

B32　Answers

1　There were thirty-five apples in a box. Thirteen are sold.
　How many apples remain?　22
2　Write in figures: one thousand and four.　1004
3　Divide nine by three.　3
4　A fourteen-metre sand pit has eight metres added to it, but is then reduced by three metres.
　What is the final length?　19 m
5　There are fourteen people on a bus. Seven get off and four get on.
　How many people are now on the bus?　11
6　Multiply eighty-four by ten.　840
7　How many fifths are there, in two whole ones?　10
8　There are six eggs in a box.
　How many eggs are there, altogether, in seven boxes?　42
9　What is nine squared?　81
10　Divide three hundred and forty by one hundred.　3.4

11　A watch is ten minutes fast.
　When the watch shows half past eight, what is the correct time?　8.20*
12　Find the total of two ten-pence coins, and five five-pence coins.　45p
13　Subtract twenty-one from thirty-seven.　16
14　What is twelve less than forty?　28
15　A video recorder priced at two hundred and ninety-nine pounds is reduced by seventy-nine pounds.
　What is its reduced price?　£220
16　A table is to be sold with four chairs.
　How many chairs will be needed for sale with thirty-seven tables?　148
17　To five point two, add three point nought two.　8.22
18　Write down the length of the perimeter of a square of length six centimetres.　24 cm
19　Subtract forty-two from one hundred and twenty-nine.　87
20　Eighty-eight per cent of a batch of mice were white.
　What percentage of the group were a colour other than white?　12%

21　Two boxes, one of height thirty-seven centimetres and the other seventy-six centimetres, are stacked on top of each other.
　What is the total height of the stack, in centimetres?　113 cm
22　Multiply twenty-six by eleven.　286
23　Subtract twenty-eight from forty-five.　17
24　How many twelves are there, in four hundred and eighty?　40
25　A newspaper delivery van has a route which takes three and three quarters of an hour.
　The last paper can be delivered no later than half past four in the afternoon.
　What is the latest time, the van can start delivering?　12.45*

37

B33

1. Sally uses ten pints of milk a week. How many pints will she use in seven weeks? — 70 pints
2. What is eight less than seven hundred and six? — 698
3. What is the greatest number of five-pence coins you could get from thirty-eight pence? — 7
4. Add twenty-three and fifty-two. — 75
5. Subtract thirty-three from fifty. — 17
6. Write down a multiple of eight, which is less than twenty. — 8 or 16
7. How many tens are there, in seven hundred and seventy? — 77
8. A man takes seven hours to make a wooden vase. How long will it take him to make nine vases? — 63 hrs
9. What is six squared? — 36
10. Multiply sixty-five by one hundred. — 6500

11. What is the combined weight of forty-three grams and fifty seven grams? — 100 g
12. Subtract seventeen from forty-two. — 25
13. Add together nineteen, twenty-one, and twenty-three. — 63
14. What is the time, fifteen minutes after ten fifty? — 11.05*
15. How much change do you get from ten pounds, if you spend three pounds and fifty pence? — £6.50
16. Multiply fifteen by six. — 90
17. To nought point one four, add nought point two five. — 0.39
18. On one particular day, the rain started at one thirty-five, and lasted for forty minutes. At what time, did the rain stop? — 2.15*
19. From one hundred and twenty-one, subtract thirty-seven. — 84
20. One eighth of a number is twelve. What is the number? — 96

21. Add seven thousand, four hundred and fifty millimetres, and six hundred millimetres. — 8050 mm
22. What is twice twenty-eight? — 56
23. How many elevens are there, in six hundred and sixty? — 60
24. Multiply twenty-five by twelve. — 300
25. Add together forty-eight and fifty-four. — 102

B34

1. How many hours are there, between oh two fifteen hours and seventeen fifteen hours? — 15 hrs
2. What is nineteen less than twenty-three? — 4
3. How many fives are there, in thirty? — 6
4. What is seven hundred and forty-five, to the nearest hundred? — 700
5. There are sixty-six pupils going on a French trip. Twenty-two more pupils then want to go. How many will there be, altogether? — 88
6. Divide nine hundred and seventy-four by one hundred. — 9.74
7. Multiply six by eight. — 48
8. Two out of every ten people are in a year group do not like chips. Write this as a fraction. — $\frac{2}{10}$ or $\frac{1}{5}$
9. How many sevenths are there, in five whole ones? — 35
10. A box contains ten tubes of toothpaste. How many tubes will there be in seventy-six boxes? — 760

11. Eighty-two parents at a school football match are joined by eighteen from the opposite side. How many parents are there, altogether? — 100
12. Add seventy-two and sixty-six. — 138
13. Subtract twenty-three from forty-nine. — 26
14. What is ten more than four hundred and ninety-five? — 505
15. Clare needs twelve pounds for a CD, but has saved only eleven pounds and forty-five pence. How much more does she need? — 55p
16. What is twice sixty-four? — 128
17. From seven, subtract five point five. — 1.5
18. A test started at five to eleven, and finished at half past eleven. How long was the test? — 35 min
19. Multiply fifty-four by five. — 270
20. Apples are sold in bags of four each. Thirty-five apples are needed. How many bags will need to be bought? — 9

21. Seventy-two pairs of twins attend a gathering in London. How many people are there, altogether? — 144
22. Multiply thirty-five by twelve. — 420
23. How much change do you get from twenty-five pounds, if you buy three calculators each costing six pounds and eighty pence? — £4.60
24. From sixty-three, subtract forty-nine. — 14
25. How many nineteens are there, in three hundred and eighty. — 20

B35

1 Sixteen sweets are divided equally between two children.
How many does each one receive? — 8

2 Add together four, three, seven, and two. — 16

3 What are ten sixes? — 60

4 A tank contains fourteen litres.
Three litres are drawn off and, then, ten litres are added.
How many litres are now in the tank? — 21 litres

5 Add twenty-eight to twelve. — 40

6 How many tens are there, in two thousand, four hundred? — 240

7 What is three squared? — 9

8 Forty-two days is the same as how many weeks? — 6 wks

9 Multiply nine by six. — 54

10 It will cost fourteen pounds to send a box to America.
How much will it cost to send one hundred similar boxes to America? — £1400

11 A hi-fi unit is reduced, from two hundred and fifty pounds to one hundred and ninety-nine pounds.
By how much has it been reduced? — £51

12 Add together eighteen, twenty, and twenty-two. — 60

13 What is twice forty-five? — 90

14 Add sixty-nine to twenty-two. — 91

15 Find the total of five ten-pence coins and four five-pence coins. — 70p

16 Add forty-three to forty-three. — 86

17 From six point nine, subtract two point four. — 4.5

18 How many halves are there, in six and a half whole ones? — 13

19 Multiply sixty-three by eight. — 504

20 Subtract fifty-two from one hundred and seven. — 55

21 A large house needs twenty-nine pairs of curtains.
How many curtains is this, altogether? — 58

22 How many thirteens are there, in six hundred and fifty? — 50

23 From eighty-four, subtract thirty-seven. — 47

24 Andy works as a part-time gardener each day from half past eleven until quarter past four.
How long, in hours and minutes, does he work each day? — 4 hrs 45 min

25 Multiply fifty-six by eleven. — 616

B36

1 What is the greatest number of two-pence coins you could get from twenty-three pence? — 11

2 From forty, subtract twenty-eight. — 12

3 How many fours are there, in eight? — 2

4 There are eighty-seven people in a hotel.
Six leave and, then, eight new people register to stay. How many will there now be, in the hotel? — 89

5 What is thirty-two, added to seventeen? — 49

6 A shop has ninety-three bundles of balsa wood. There are ten pieces in each bundle. How many pieces of wood are there, altogether? — 930

7 What is eight squared? — 64

8 Divide two hundred and twenty-seven by one hundred. — 2.27

9 What is two times two times two? — 8

10 Six children are each aged nine years of age. What is their total combined age? — 54 yrs

11 A clock is twenty minutes slow.
When the clock shows twenty past four, what is the correct time? — 4.40*

12 What is twelve less than fifty? — 38

13 Add eighteen and seventy-six. — 94

14 How much change do you get from five pounds, if you spend two pounds and eighty pence? — £2.20

15 Add thirty-two and sixty-eight. — 100

16 One seventh of a number is twelve.
What is the number? — 84

17 Multiply sixty-one by nine. — 549

18 To two point seven one, add three point nought four. — 5.75

19 Simon left school at half past three, and arrived home at five minutes past four. How long did it take him to get home? — 35 min

20 From one hundred and seventeen, subtract fifty-eight. — 59

21 Find the total of three ten-pence coins, four five-pence coins, and five two-pence coins. — 60p

22 Add seventy-two and sixty-seven. — 139

23 Add two thousand, nine hundred and forty pounds, and two hundred pounds. — £3140

24 Multiply sixty-four by eleven. — 704

25 You buy four train tickets costing six pounds and fifty pence each.
What change should you receive from thirty pounds? — £4

1. Eighteen crayons are divided equally between two children.
   How many crayons will each child receive? **9**
2. Subtract sixteen from twenty-three. **7**
3. Add fourteen and twenty-three. **37**
4. What are six fives? **30**
5. Three pencils are laid end to end.
   The pencils are of lengths twelve centimetres, eight centimetres, and five centimetres.
   What will be the total length? **25 cm**
6. Multiply nine by seven. **63**
7. How many sixths are there, in four whole ones? **24**
8. Write down a multiple of six less than twenty. **6, 12 or 18**
9. How many hundreds are there, in nine thousand? **90**
10. A box contains sixty mints.
    How many mints will there be, altogether, in ten boxes? **600**

11. A set of videos costing sixty-six pounds is halved in price.
    What is the new price? **£33**
12. What is ten more than eight hundred and ninety-eight? **908**
13. What is the time, twenty minutes before five past one? **12.45***
14. Add nineteen and fourteen. **33**
15. Subtract twenty-seven from forty-two. **15**
16. What is twice forty-one? **82**
17. From four, subtract two point four. **1.6**
18. Of a hundred television sets tested, fourteen had a fault.
    Write this as a percentage. **14%**
19. What is fifty-two, multiplied by seven? **364**
20. A forty-five minute game finished at twenty past two.
    At what time, did it start? **1.35***

21. Sandy buys three board games, each costing nine pounds and forty pence.
    What change should she receive from thirty pounds? **£1.80**
22. From nine, subtract five point seven five. **3.25**
23. What is thirty-six less than fifty-three? **17**
24. Add forty-four to fifty-nine. **103**
25. What is twenty-two, multiplied by fourteen? **308**

1. The date today is the fourteenth.
   What will be the date in exactly one week's time? **21st**
2. What is five hundred and fifty-five, to the nearest hundred? **600**
3. From thirty-three, subtract twenty-seven. **6**
4. Add together two, five, four, and nine. **20**
5. Add thirty-two and sixteen. **48**
6. What is two squared? **4**
7. A dozen is twelve.
   How many will ten dozen be? **120**
8. Divide six hundred and seventy by one hundred. **6.7**
9. What is six, multiplied by seven? **42**
10. Write down a factor of eighteen. **1, 2, 3, 6, 9 or 18**

11. How much change should you get from ten pounds, if you spend five pounds and fifty pence? **£4.50**
12. Find the total of four ten-pence coins, and four five-pence coins. **60p**
13. Add fifty-seven and twenty-six. **83**
14. What is sixteen less than forty-three? **27**
15. Add together eighteen, nineteen, and twenty. **57**
16. To five point one two, add two point two one. **7.33**
17. How many quarters are there, in twelve whole ones? **48**
18. There are fifty-two weeks in one year.
    How many weeks are there, in four years? **208 wks**
19. From one hundred and thirteen, subtract thirty-four. **79**
20. Simon started a cleaning job at half past eleven, and finished at ten past twelve.
    How long did the job take? **40 min**

21. A machine part, which weighs seven thousand, seven hundred and seventy grams, is packed into a box which weighs four hundred grams.
    What is the total weight? **8170 g**
22. What is twice thirty-nine? **78**
23. How many elevens are there, in three hundred and thirty? **30**
24. Multiply seventeen by twelve. **204**
25. Four books are bought which each cost six pounds and fifty pence.
    What change should you receive from thirty pounds? **£4.00**

1 How many hours are there, between ten hundred hours and twenty hundred hours?　　　　　　　　　　10

2 How many twos are there, in twenty-two?　　11

3 From sixty, subtract forty-six.　　　　14

4 A man drives fifteen kilometres away from home, turns around and drives six kilometres back towards his home and, then, drives fifteen kilometres further away from home.
How far is he now from his home?　　24 km

5 What are ten tens?　　　　　　　100

6 Write down a multiple of five, which is less than twenty.　　　　5, 10 or 15

7 What is twenty-seven, multiplied by one hundred?　　　　　　　2700

8 How many days are there, in eight weeks?　　　　　　　　56 days

9 Divide five hundred and ninety-eight by ten.　　　　　　　　59.8

10 How many ninths are there, in five whole ones?　　　　　　　45

11 Border plants are sold as a tray of nine plants. How many trays will be needed by a gardener who wants seventy border plants?　　　　　　　　8

12 Subtract twenty-six from sixty.　　34

13 How many seconds are there, in twenty minutes?　　　　　　1200 sec

14 Add thirty-three and sixty-seven.　100

15 What is ten more than two hundred and ninety-nine?　　　　309

16 Add twenty-four and thirty-eight.　62

17 Subtract forty-four from one hundred and seven.　　　　　　　63

18 What is eleven, multiplied by nine?　99

19 To nought point one two, add nought point two one.　　　　　0.33

20 A fifty-minute coach journey started at twenty-five minutes to eleven.
At what time, did it finish?　　11.25*

21 A woman won a lottery prize of four thousand, five hundred and fifteen pounds, and a second prize of six hundred pounds.
How much did she win, altogether?　£5115

22 Add eighty-nine and forty-two.　131

23 How many sixteens are there, in eight hundred?　　　　　　　50

24 Multiply fifty-eight by eleven.　638

25 What change should be received from a fifty-pound note, when four concert tickets are bought at eleven pounds, fifty pence each?　　　　£4

1 Twelve fairy cakes are divided equally between three children.
How many cakes does each child receive?　　4

2 What are five tens?　　　　　50

3 The scores on two dice are six and five.
What is the total score?　　　11

4 What is three less than two hundred?　197

5 Add fourteen and twenty-four.　38

6 How many tens are there, in two hundred and ten?　　　　21

7 What is five squared?　　　　25

8 Fifty-six days is the same as how many weeks?　　　　　　8 wks

9 What are six eights?　　　　48

10 Multiply forty-eight by one hundred.　4800

11 One machine tool weighs one hundred and seventy-five grams.
What will be the weight of two such machine tools?　　　　350 g

12 Add together twenty-three, twenty-one, and nineteen.　　　　　63

13 Find the total of five ten-pence coins and five five-pence coins.　　75p

14 Add seventy-three and twenty-seven.　100

15 Find the total of twenty-eight pounds and sixty-four pounds.　　　£92

16 From four, subtract one point five.　2.5

17 How many sixths are there, in six whole ones?　　　　　　36

18 What is forty-seven, multiplied by five?　235

19 What is twice forty-four?　　88

20 In a school, sixty-four per cent of pupils like sports lessons.
What percentage of children dislike sports lessons?　　　　36%

21 A tank contains one thousand, five hundred and fifteen litres of water.
Eight hundred litres are added.
How much water is there now, in the tank?　　　　2315 litres

22 Add sixty-seven and thirty-eight.　105

23 How many seventeens are there, in eight hundred and fifty?　　50

24 A paving job will take four and a quarter hours to complete. If you want to be finished by half past three, at what time should you start?　　　11.15*

25 Multiply seventeen by thirteen.　221

C1

1. There are ten exercise books in a pack. How many exercise books are there, in twenty-three packs? — **230**
2. Multiply four by seven. — **28**
3. How many hundreds are there, in three thousand, five hundred? — **35**
4. Write down a factor of fourteen. — **1, 2, 7 or 14**
5. What is eight squared? — **64**
6. Subtract seven from four. — **−3**
7. A device needs five fuses. How many fuses are needed for sixty-seven devices? — **335**
8. Multiply three thousand by eighty. — **240000**
9. How many metres are there, in two hundred and twenty-four centimetres? — **2.24 m**
10. Approximately, how many pounds weight are there, in ten kilograms? — **22 lb**

11. Sixty-three per cent of the workers at a factory are male. What percentage are female? — **37%**
12. How many eighths are there, in sixteen whole ones? — **128**
13. A ruler costs five pence. How much will fifty-eight rulers cost? — **£2.90**
14. To nought point nought three, add nought point one two. — **0.15**
15. Look at your answer sheet. Write down an estimated reading from the dial. — **(2.8 kg)**
16. Find one eighth of ninety-six kilometres. — **12 km**
17. What is seventy per cent of two hundred and fifty pounds? — **£175**
18. A lorry is carrying a load of seven hundred and four kilograms. A box of weight forty-seven is removed. What weight is left? — **657 kg**
19. Multiply two hundred and twelve by seven. — **1484**
20. Look at your answer sheet. Estimate the size of the angle in degrees. — **30°**

21. There are sixteen chocolates in a bag. How many chocolates are there, in a dozen bags? — **192**
22. From six, subtract four point two five. — **1.75**
23. How many twelves are there, in two hundred and forty? — **20**
24. A salesman drives six hundred and fifty kilometres on Monday and, then, four hundred and sixty-five kilometres on Tuesday. What is the total distance, he has driven? — **1115 km**
25. Divide three thousand, two hundred and seventy by six. — **545**

C2

1. There are three leaves on a clover. How many leaves are there, altogether, on eight clovers? — **24**
2. How many tens are there, in two hundred and thirty? — **23**
3. Multiply nineteen by one hundred. — **1900**
4. How many weeks are there, in sixty-three days? — **9 wks**
5. How many centimetres are there, in thirty-three millimetres? — **3.3 cm**
6. One person in ten does not have a television. Write this as a fraction. — **$\frac{1}{10}$**
7. Multiply forty-two by six. — **252**
8. How many kilograms are there, in six thousand grams? — **6 kg**
9. Multiply three hundred by fifty. — **15000**
10. A temperature of minus two degrees rises by seven degrees. What is the new temperature? — **5°**

11. A television programme starts at twenty past two, and lasts for fifty minutes. At what time does it finish? — **3.10***
12. From one hundred and twenty-seven, subtract fifty-nine. — **68**
13. From eight, subtract four point two. — **3.8**
14. What are twenty-nine sixes? — **174**
15. Write down the length of the perimeter of a square of side seven centimetres. — **28 cm**
16. What is two hundred and eighty-seven, add thirty-six? — **323**
17. Look at your answer sheet. Write down an estimate for the length of the line, in centimetres. — **(2 cm)**
18. Write down all the prime numbers between ten and fifteen. — **11, 13**
19. What is one hundred and twenty-three, multiplied by six? — **738**
20. What is three eighths of forty-eight hours? — **18 hrs**

21. Becky buys three trays of cat food at a discount warehouse for five pounds and fifty pence each. What change should she receive from a twenty-pound note? — **£3.50**
22. Add four hundred and fifty-six, to eight hundred and fifty-four. — **1310**
23. How many sixteens are there, in nine hundred and sixty? — **60**
24. What is one thousand, three hundred and eighty-eight, divided by four? — **347**
25. Multiply twenty-two by eleven. — **242**

C3

1 Apples are sold four to a packet.
How many packets can be made up from
a box of thirty-six apples?   9

2 How many hundreds are there, in seven
hundred?   7

3 What is two times two times two?   8

4 What are seven eights?   56

5 Write down a multiple of four, which is
less than twenty.   4, 8, 12 or 16

6 A temperature is minus four degrees.
The temperature falls by three degrees.
What is the new temperature?   −7°

7 How many centimetres are there, in
three point eight metres?   380 cm

8 A lottery has sixty prizes, each of five
thousand pounds.
What is the total prize money?   £300,000

9 Multiply thirty-six by seven.   252

10 How many litres are there, in three
thousand, six hundred and thirty
millilitres?   3.63 litres

11 There are twenty-four packets of crisps
in an assorted bag.
How many packets are there, in eight
bags?   192

12 One seventh of a number is eleven.
What is the number?   77

13 What is twice forty-three?   86

14 From seven point four seven, subtract
nought point one three.   7.34

15 Look at your answer sheet.
Estimate the size of the angle, in degrees.   (35°)

16 What is fifty-five less than three hundred
and three?   248

17 How many minutes are there, in one
and three quarters of an hour?   105 min

18 What is half of twenty-three yards? $11\frac{1}{2}$ yds

19 Multiply three hundred and twenty-one
by five.   1605

20 Twenty-five per cent of a number is
eighty. What is the number?   320

21 Humad makes three telephone
calls, each costing five pounds and
fifty pence.
How much will be left out of a
twenty-pound budget for the calls?   £3.50

22 How many fifteens are there, in four
hundred and fifty?   30

23 What number is halfway between four
point five and ten point five?   7.5

24 What is four thousand, two hundred
and eight, divided by eight?   526

25 In one box, there are three hundred
and twenty-seven nails.
In a second box, there are eight hundred
and ninety-six nails.
How many nails are there, altogether?   1223

C4

1 Exam tables are arranged eight to a row.
How many rows are there, when all
sixty-four tables are set out?   8

2 How many pence are there, in fifty-four
pounds?   5400p

3 A half-dozen is six.
How many is four half-dozens?   24

4 How many tens are there, in one
hundred and ninety?   19

5 How many tenths are there, in ten
whole ones?   100

6 Multiply six hundred by three hundred.   180000

7 How many centimetres are there, in
nine point eight metres?   980 cm

8 From two, subtract six.   −4

9 Write in figures: half a million.   500000

10 How many kilometres are there, in
five thousand, seven hundred and
fifty metres?   5.75 km

11 Jeremy takes twenty minutes to eat his
tea, starting at five to six.
At what time does he finish?   6.15*

12 How many halves are there, in seven
and a half whole ones?   15

13 Multiply forty-two by four.   168

14 From nine point one four, subtract
seven point nought three.   2.11

15 Twenty-four per cent of the rabbits in
a zoo are brown.
What percentage are a colour other
than brown?   76%

16 On the line AB on your answer sheet,
put an X at a point which you estimate
to be thirty millimetres from A.   (30 mm)

17 Add sixty-eight to eight hundred and
ninety-three.   961

18 A block of cheese weighs three hundred
and fifty grams.
Thirty per cent is cut off.
What weight of cheese is cut off?   105 g

19 What is three quarters of forty-eight
hours?   36 hrs

20 Five hundred and twelve people pay
five pounds each to see a film.
What is the total amount paid?   £2560

21 What is three thousand, four hundred
and sixty-five, divided by
nine?   385

22 Louise has two hundred and thirty-six
stamps. Emma has eight hundred and
eighty-six stamps. How many stamps
have they, altogether?   1122

23 How many twelves are there, in eight
hundred and forty?   70

24 What number is halfway between two
point three and four point three?   3.3

25 Look at the map on your answer sheet,
which has a scale of one centimetre to
five kilometres.
Estimate the direct distance from point
A to point B, in kilometres.   (10 km)

C5

1 On four days during a particular week, the rain fell.
During what fraction of the week was there dry weather? — $\frac{3}{7}$

2 How many tens are there, in four hundred and thirty? — 43

3 What is three squared? — 9

4 What are nine sixes? — 54

5 How many sixths are there, in three whole ones? — 18

6 What is minus three, added to minus seven? — −10

7 How many millimetres are there, in seven and a half centimetres? — 75 mm

8 A coach can carry forty-eight people. How many people can be carried by eight of these coaches? — 384

9 Multiply eighty by forty. — 3200

10 How many kilograms are there, in thirty-eight thousand grams? — 38 kg

11 A radio programme runs from quarter to seven to seven twenty-five. How long is the programme? — 40 min

12 What is twice eighty-four? — 168

13 To two point two, add one point eight. — 4

14 Look at your answer sheet. Write down an estimate of the reading from the dial. — (2.33 m)

15 What are fifty-three sixes? — 318

16 What is forty-five less than two hundred and twenty-four? — 179

17 Twenty-five per cent of eight hundred buildings are for office use. How many buildings is this? — 200

18 How many minutes are there, in one and a third hours? — 80 min

19 What is nine tenths of four hundred? — 360

20 On your answer sheet, draw an arrow on the protractor scale to indicate your estimate of a forty-five degree angle. — (45°)

21 To eight hundred and eighty-four, add three hundred and fifty-six. — 1240

22 Divide three thousand, two hundred and fifty-five by seven. — 465

23 One dozen is twelve. How many is fourteen dozen? — 168

24 What is six, minus four point two five? — 1.75

25 How many elevens are there, in three hundred and thirty? — 30

C6

1 How many metres are there, in seven hundred and sixty centimetres? — 7.6 m

2 How many weeks are there, in forty-nine days? — 7 wks

3 What is seven hundred and thirty, multiplied by ten? — 7300

4 How many sevenths are there, in four whole ones? — 28

5 Multiply four thousand by seventy. — 280000

6 How many hundreds are there, in five thousand, five hundred? — 55

7 What are three threes? — 9

8 Wilson needs to take nine blood pressure tablets each day. How many will he take over a 22-day period? — 198

9 A temperature falls by seven degrees from minus three degrees. What is the new temperature? — −10°

10 How many millilitres are there, in eight point one litres? — 8100 ml

11 A minibus can carry eight passengers. How many minibuses will be needed for thirty-five passengers? — 5

12 To five point nought one, add two point two. — 7.21

13 What is thirty-one less than one hundred and twenty-six? — 95

14 Multiply forty-six by three. — 138

15 What is the greatest number of twenty-pence coins you could get for eleven pounds? — 55

16 One fifth of a number is forty. What is the number? — 200

17 From seven hundred and nine, subtract fifty-four. — 655

18 A man pays five per cent of his earnings into a pension fund. How much does he pay in from ninety pounds earnings? — £4.50

19 Multiply four hundred and four by nine. — 3636

20 What is two hundred and ten minutes, written in hours? — $3\frac{1}{2}$ hrs

21 What number is halfway between six point five and twelve point five? — 9.5

22 How many thirteens are there, in three hundred and ninety? — 30

23 A school is open from half past eight to five past three. How long, in hours and minutes, is the school open? — 6 hrs 35 min

24 On a Thursday, eight hundred and seventy-two apples are picked but, on Friday, only two hundred and forty-eight are picked. How many apples have been picked, altogether? — 1120

25 How many eights are there, in five thousand, nine hundred and seventy-six? — 747

C7

		Answers
1	A single glass costs seven pounds. How much will eight glasses cost?	£56
2	What is fourteen, multiplied by one hundred?	1400
3	Write down a multiple of three, which is less than ten.	3, 6 or 9
4	What is three times three times three?	27
5	How many tens are there, in eight thousand, four hundred?	840
6	From minus four, subtract minus two.	−2
7	How many centimetres are there, in thirty millimetres?	3 cm
8	Multiply four hundred by three thousand.	120000
9	Three planks, each of width seventy-eight millimetres, are placed on top of each other. What is their total width?	234 mm
10	How many grams are there, in thirty kilograms?	30000 g
11	What number is twenty-eight less than one hundred and twenty-five?	97
12	What is seven minus five point eight?	1.2
13	How many thirds are there, in nine whole ones?	27
14	In a car, there are five men who are each forty-four years of age. What is their total age?	220 yrs
15	Look at your answer sheet. Estimate the size of the angle, in degrees.	(60°)
16	What is one fifth of one hundred and twenty gallons?	24 gal
17	From four hundred and eighty-three, subtract seventy-six.	407
18	Fifty per cent of forty-six plants have died. How many plants have died?	23
19	Multiply one hundred and twenty-three by four.	492
20	Write three and a quarter hours, in minutes.	195 min
21	How many sevens are there, in two thousand, four hundred and ninety-two?	356
22	Multiply forty-six by eleven.	506
23	Look at the map on your answer sheet, which has a scale of one centimetre to four kilometres. Estimate the direct distance from point A to point B in kilometres.	(10 km)
24	How many twelves are there, in four hundred and eighty?	40
25	Add two hundred and ninety-four, and nine hundred and thirty-seven.	1231

C8

		Answers
1	How many six-seater minibuses will be needed for forty-eight people?	8
2	Write down a factor of fifteen.	1, 3, 5 or 15
3	What is five hundred and thirty, divided by one hundred?	5.3
4	What are seven fours?	28
5	What is twenty-two, multiplied by ten?	220
6	How many litres are there, in four thousand and ninety millilitres.	4.09 litres
7	Multiply three hundred by forty.	12000
8	From three, subtract nine.	−6
9	How many metres are there, in six point eight kilometres?	6800 m
10	Approximately, how many inches are equivalent to ten centimetres?	4 in
11	Of one hundred people on a ferry, only nine felt seasick. Write this as a percentage.	9%
12	To five point three, add two point four.	7.7
13	What is twenty-two, multiplied by eight?	176
14	Look at your answer sheet. Write down an estimate of the reading on the dial.	(6.7 amps)
15	What is sixty-seven less than five hundred and four?	437
16	A twenty-minute test started at quarter to eleven. At what time did the test finish?	11.05*
17	Ten per cent of a number is thirty. What is the number?	300
18	What is five eighths of seventy-two millilitres?	45 ml
19	Look at your answer sheet. Estimate the length of the line, in millimetres.	(40 mm)
20	What is three hundred and thirteen, multiplied by four?	1252
21	What number is halfway between six point five and eight point five?	7.5
22	How many twelves are there in six hundred?	50
23	Find the change from twenty-five pounds, when three shirts are bought each costing six pounds and fifty pence.	£5.50
24	How many sixes are there, in four thousand and sixty-eight?	678
25	Add five hundred and forty-six to six hundred and seventy-five.	1221

1 Three cats each have four kittens. How many kittens are there, altogether?   12

2 How many eighths are there, in five whole ones?   40

3 Divide two hundred and fifty by one hundred.   2.5

4 What is sixty-two, multiplied by ten?   620

5 A box contains nine tennis balls. How many boxes will be needed for forty-five tennis balls?   5

6 How many centimetres are there, in three and a half metres?   350 cm

7 Multiply fifty-four by four.   216

8 A temperature of five degrees falls by eight degrees. What is the new temperature?   –3°

9 What is three hundred, multiplied by thirty?   9000

10 How many kilograms are there, in nine thousand, six hundred and sixty grams?   9.66 kg

11 Write down the length of the perimeter of a square of side eight centimetres.   32 cm

12 What is twice sixty-two?   124

13 To nought point three three, add two point three.   2.63

14 Lee has one hundred and thirteen pounds. He spends forty-six pounds on a new jacket. How much money is left?   £67

15 Multiply thirty-six by six.   216

16 Find twenty per cent of two hundred and ten grams.   42 g

17 Add three hundred and ninety-seven, and forty-nine.   446

18 Jules plans to spend a third of his three thousand, six hundred lire each day of a three-day break. What amount should he spend each day?   1200 lire

19 Write one and a quarter hours, in minutes.   75 min

20 On your answer sheet, draw an arrow on the protractor scale to indicate your estimate of a thirty-degree angle.   (30°)

21 Find the amount remaining in a one hundred-pound budget, when four CDs are bought, each costing fifteen pounds and fifty pence.   £38

22 Multiply seventy-six by eleven.   836

23 How many thirteens are there, in two hundred and sixty?   20

24 Six thousand and thirty-two plastic ducks are divided equally and put into eight sacks. How many ducks will there be, in each sack?   754

25 Add four hundred and fifty-nine to seven hundred and fifty-two.   1211

1 CDs are being sold for nine pounds each. What is the total cost of eight CDs?   £72

2 Divide forty-eight by ten.   4.8

3 What is four squared?   16

4 How many sixes are there, in fifty-four?   9

5 Multiply twenty-seven by one hundred.   2700

6 How many millimetres are there, in eight point nine centimetres?   89 mm

7 Fifty people on a coach trip have each brought twenty pounds spending money. What is the total amount brought by those on the coach?   £1000

8 A temperature of minus four rises by nine degrees. What is the new temperature?   5°

9 How many litres are there, in seven thousand, five hundred millilitres?   7.5 litres

10 Multiply ninety-five by five.   475

11 A cross channel ferry left Dover at ten to twelve, and arrived in Calais at half past twelve. How long did the crossing take?   40 min

12 From three, subtract one point one.   1.9

13 One seventh of a number is eight. What is the number?   56

14 What is twice fifty-six?   112

15 Multiply sixty-four by seven.   448

16 Wendy has spent seventy-five per cent of eight thousand lire spending money while on holiday. What is the actual amount, she has spent?   6000 lire

17 Add fifty-five to seven hundred and eighty-nine.   844

18 Multiply two hundred and fourteen by three.   642

19 What is two thirds of twenty-seven?   18

20 On your answer sheet, draw an arrow on the protractor scale to indicate your estimate of a sixty-degree angle.   (60°)

21 A tank containing eight hundred and eighty-eight litres has a further four hundred and seventeen litres added. How many litres are now in the tank?   1305 litres

22 How many sixteens are there, in four hundred and eighty?   30

23 Multiply eleven by fifteen.   165

24 A ferry left port at eleven fifty hours, and returned at fifteen twenty-five hours. How long, in hours and minutes, was the journey?   3 hrs 35 min

25 Divide three thousand, one hundred and eighty-four by four.   796

C11
Answers

1. A doctor can see eight patients each hour. How many hours will it take for the doctor to see forty-eight patients? — 6 hrs
2. Divide seven hundred and forty by one hundred. — 7.4
3. Write down a multiple of ten, which is less than thirty. — 10 or 20
4. Multiply sixty-three by ten. — 630
5. How many days are there, in three weeks? — 21
6. From two, subtract four. — −2
7. How many grams are there, in seven point three kilograms? — 7300 g
8. Multiply seven thousand by two thousand. — 14000000
9. Approximately, how many centimetres are equivalent to ten feet? — 300 cm
10. How many kilometres are there, in thirty thousand metres? — 30 km

11. Three hundred and three people attended a parents' evening, but thirty-seven arrived late. How many people arrived on time? — 266
12. What is thirty-three multiplied by five? — 165
13. How many quarters are there, in eight whole ones? — 32
14. What is forty-nine less than one hundred and four? — 55
15. Look at your answer sheet. Write down an estimated reading from the dial. — (12.2°C)
16. From five point two four, subtract two point one two. — 3.12
17. What is half of thirty-nine grams? — 19.5 g
18. Twenty per cent of a number is twenty. What is the number? — 100
19. Multiply four hundred and thirteen by five. — 2065
20. What is two hundred and seventy minutes, written in hours? — $4\frac{1}{2}$ hrs

21. Add seven hundred and eighty-eight to five hundred and twenty-six. — 1314
22. How many fourteens are there, in five hundred and sixty? — 40
23. Divide four thousand and eighty-six by nine. — 454
24. What number is halfway between four point five and twelve point five? — 8.5
25. Look at the map on your answer sheet, which has a scale of one centimetre to ten kilometres. Estimate the direct distance from point A to point B, in kilometres. — (23 km)

C12
Answers

1. A journey takes forty-two days. How many weeks are there, in forty-two days? — 6 wks
2. Multiply sixty-seven by one hundred. — 6700
3. What is four squared? — 16
4. Divide fifty-five by ten. — 5.5
5. How many sixths are there, in three whole ones? — 18
6. How many centimetres are there, in thirty-seven millimetres? — 3.7 cm
7. A temperature of minus three degrees falls by six degrees. What is the new temperature? — −9°
8. A brief case costs thirty-four pounds. What will be the total cost of buying six brief cases? — £204
9. Multiply ninety by three thousand. — 270000
10. How many litres are there, in two thousand, nine hundred and ten millilitres? — 2.91 litres

11. The snow started at quarter to twelve, and lasted for thirty-five minutes. At what time did it stop snowing? — 12.20*
12. What is a half, written as a percentage? — 50%
13. Add four point seven and two point one three. — 6.83
14. What is one eighth of four hundred and eighty? — 60
15. Multiply sixteen by nine. — 144
16. Write down all the prime numbers between fifteen and twenty. — 17, 19
17. A hospital doctor can see nine people in a day. How many whole days will it take for the doctor to see fifty people? — 6 days
18. What is eighty-nine less than two hundred and eighteen? — 129
19. Write three and three quarters of an hour, in minutes. — 225 min
20. On the line AB on your answer sheet, put an X at a point which you estimate to be two centimetres from A. — (2 cm)

21. A plane left an airport at ten fifty hours for a four and a quarter hour flight. At what time did the plane land? — 15.05*
22. How many twelves are there, in three hundred and sixty? — 30
23. Multiply thirty-four by eleven. — 374
24. Add six hundred and eighty-six to five hundred and thirty-five. — 1221
25. How many sixes are there, in two thousand, two hundred and fifty? — 375

		Answers
1	There are six eggs in one half-dozen box. How many eggs are there in seven half-dozen boxes?	42
2	How many fifths are there, in four whole ones?	20
3	Multiply twenty-four by one hundred.	2400
4	How many tens are there, in five hundred and fifty?	55
5	A taxi can carry four people. How many taxis are needed for twenty-four people?	6
6	How many centimetres are there, in seven metres?	700 cm
7	Multiply two hundred by fifty.	10000
8	Write in figures: one quarter of a million.	250000
9	To minus four, add minus three.	−7
10	How many kilometres are there, in four thousand metres?	4 km

11	Thirty-nine out of a hundred people taking a driving test had taken the test before. For what percentage of people was this their first driving test?	61%
12	What is twice fifty-one?	102
13	There are fifty-two cards in a pack. How many cards are there, in eight packs?	416
14	From one hundred and nine, take forty-three.	66
15	Look at your answer sheet. Estimate the size of the angle, in degrees.	(45°)
16	Find twenty per cent of nine hundred and fifty grams.	190 g
17	How many is a quarter of ninety-six people?	24
18	What is three hundred and twelve, multiplied by six?	1872
19	From five hundred and seven, take sixty-four.	443
20	Write one hundred and thirty-five minutes, in hours.	$2\frac{1}{4}$ hrs

21	Three tins of baby milk are bought, at a cost of five pounds and fifty pence each. How much change should you receive out of twenty pounds?	£3.50
22	How many fourteens are there, in two hundred and eighty?	20
23	Multiply sixty-six by eleven.	726
24	Divide four thousand, one hundred and four by nine.	456
25	Add eight hundred and forty-two, and four hundred and ninety-eight.	1340

		Answers
1	There are one hundred and ten bricks in a box of toy building bricks. How many bricks are there, in ten boxes?	1100
2	What is six squared?	36
3	Multiply eight by seven.	56
4	Divide nine hundred and twenty-four by one hundred.	9.24
5	Seven out of ten people have been on a train at some time. Write this as a fraction.	$\frac{7}{10}$
6	Take nine from seven.	−2
7	How many metres are there, in one hundred and ninety centimetres?	1.9 m
8	Multiply two thousand by twenty.	40000
9	A railway carriage can carry eighty-four people. What is the maximum number of people that can be carried by a train of seven carriages?	588
10	How many grams are there, in three point four kilograms?	3400 g

11	Jill has four hundred and thirty-three pounds, but spends seventy-five pounds on a new portable disc player. How much has she left?	£358
12	One seventh of a number is six. What is the number?	42
13	To three point nought four, add one point nought five.	4.09
14	Look at your answer sheet. Write down an estimated reading from the dial.	(24 kg)
15	It took Brian nine hours to paint a hotel room. How long will it take him to paint fourteen similar hotel rooms?	126 hrs
16	Multiply four hundred and fourteen by three.	1242
17	What is seventy-five per cent of four thousand pesetas?	3000 pts
11	From one hundred and three, subtract fifty-four.	49
18	Write two and three quarters of an hour, in minutes.	165 mins
20	What is three tenths of eighty dollars?	$24

21	Two hundred and twenty apples are shared equally between eleven people. How many apples does each person receive?	20
22	What number is halfway between three point five and thirteen point five?	8.5
23	Add nine hundred and forty-three, and three hundred and eighty-eight.	1331
24	Divide three thousand, eight hundred and seventy-two by four.	968
25	Look at the map on your answer sheet, which has a scale of one centimetre to five kilometres. Estimate the direct distance from point A to point B, in kilometres.	(12.5 km)

C15

Answers

1 How many eight-seater minibuses are needed for fifty-six passengers? **7**
2 Multiply sixty by one hundred. **6000**
3 There are three walnut chocolates in a box. How many walnut chocolates are there, altogether, in six boxes? **18**
4 Divide nine hundred and seventy-one by ten. **97.1**
5 Write down a factor of three. **1 or 3**
6 From minus eight, subtract minus two. **−6**
7 There are fifty-two matches in a box. How many matches are there, in eight boxes? **416**
8 How many millimetres are there, in six point two centimetres? **62 mm**
9 A machine can make seventy plastic cones in an hour. How many cones can be made over a period of one thousand hours? **70000**
10 How many litres are there, in eight thousand and ninety millilitres? **8.09 litres**

11 What is the greatest number of ten-pence coins you could get for eleven pounds? **110**
12 How many fifths are there, in ten whole ones? **50**
13 To three point one, add three point one. **6.2**
14 Multiply sixty-two by seven. **434**
15 A helicopter flight lasts for twenty minutes, starting at ten to five. At what time does the flight finish? **5.10***
16 Find thirty per cent of five hundred and fifty kilograms. **165 kg**
17 There are two hundred and ninety-four passengers on a ferry. A further forty-eight passengers join the ferry. What is the total number of passengers? **342**
18 What is three quarters of one hundred and twenty pounds? **£90**
19 Write one and a third hours, in minutes. **80 min**
20 On your answer sheet, draw an arrow on the protractor scale to indicate your estimate of an eighty-degree angle. **(80°)**

21 Add these numbers: six hundred and seventy-two, and five hundred and sixty-nine. **1241**
22 How many fifteens are there, in six hundred? **40**
23 A coach arrives at the bus station at sixteen forty hours, after a journey lasting four hours and fifty-five minutes. At what time did the journey start? **11.45***
24 Multiply forty-eight by eleven. **528**
25 Divide four thousand, six hundred and six by seven. **658**

C16

Answers

1 In a game, each player needs seven cards. How many cards are needed for four players? **28**
2 How many quarters are there, in six whole ones? **24**
3 Bill throws a six on a dice three times. What is his total score? **18**
4 Divide one hundred and twenty by one hundred. **1.2**
5 What are fifty-eight tens? **580**
6 How many millilitres are there, in seven point five litres? **7500 ml**
7 From four, subtract six. **−2**
8 Approximately how many feet length are there, in three hundred centimetres? **10 ft**
9 How many minutes are there, in seven hundred hours? **42000 min**
10 How many kilometres are there, in seventy thousand metres? **70 km**

11 Write down the length of the perimeter of a square of side of length nine centimetres. **36 cm**
12 From two, subtract nought point nine. **1.1**
13 Multiply twenty-six by five. **130**
14 What is thirty-seven less than one hundred and six? **69**
15 A woman left work at twenty to six, and arrived home at six twenty-five. How long did the journey take? **45 min**
16 Multiply four hundred and thirty by eight. **3440**
17 On a flight, a plane used twenty-five per cent of its load of eight hundred gallons. How many gallons of fuel were used? **200 gal**
18 What is seven tenths of nine thousand pesetas? **6300 pts**
19 Write forty minutes, as a fraction of an hour. **$\frac{2}{3}$**
20 A nine hundred and three centimetre length of rope is shortened by twenty-eight centimetres. What is its new length? **875 cm**

21 Add five hundred and seventy-four to six hundred and forty-seven. **1221**
22 Multiply twenty-two by thirteen. **286**
23 How many sixteens are there, in three hundred and twenty? **20**
24 How many nines are there, in six thousand, eight hundred and thirteen? **757**
25 Look at the map on your answer sheet, which has a scale of one centimetre to four kilometres. Estimate the direct distance from point A to point B, in kilometres. **(9 km)**

49

C17

1. How many boxes of six eggs each can be packed from a batch of forty-eight eggs? — 8
2. How many tens are there, in eight hundred and eighty? — 88
3. What is two times two times two? — 8
4. What is seven squared? — 49
5. Multiply forty-five by one hundred. — 4500
6. How many centimetres are there, in forty millimetres? — 4 cm
7. Multiply eight hundred by seventy. — 56000
8. There are thirty-one biscuits in a tin. How many biscuits will you get in nine similar tins? — 279
9. A temperature of five degrees falls by eight degrees. What is the new temperature? — –3°
10. How many grams are there, in nine point two kilograms? — 9200 g

11. Sixty-seven per cent of accidents happen in the home. What percentage of accidents happen outside the home? — 33%
12. How many sixths are there, in twelve whole ones? — 72
13. Multiply thirty-eight by four. — 152
14. To two point seven five, add nought point nought five. — 2.8
15. Look at your answer sheet. Write down an estimated reading from the dial. — (4.2 mm)
16. From eight hundred and seventeen, subtract fifty-eight. — 759
17. How many days are there, in two hundred and thirteen weeks? — 1491 days
18. Twenty-five per cent of a number is forty. What is the number? — 160
19. Two fifths of the two hundred tickets available for a play are unsold. How many tickets are unsold? — 80
20. On the line AB on your answer sheet, put an X at a point which you estimate to be three centimetres from A. — (3 cm)

21. What number is halfway between three point eight and five point eight? — 4.8
22. How many thirteens are there, in six hundred and fifty? — 50
23. How much would remain of a one hundred-pound budget after the purchase of three bouquets of flowers at twenty-five pounds and fifty pence each? — £23.50
24. Divide three thousand, five hundred and four by six. — 584
25. What is six hundred and sixty-three, added to five hundred and sixty-seven? — 1230

C18

Answers

1. Forty-five counters are shared equally between five children. How many counters should each child receive? — 9
2. What is nine hundred and ninety, divided by one hundred? — 9.9
3. How many halves are there, in seven whole ones? — 14
4. A rectangle has four points. How many points are there, altogether, in six rectangles? — 24
5. How much is seventy-one, multiplied by ten? — 710
6. How many litres are there, in nine thousand, eight hundred millilitres? — 9.8 litres
7. Multiply four thousand by ninety. — 360000
8. A temperature of minus three degrees rises by seven degrees. What is the new temperature? — 4°
9. There are six table settings in each box of crockery. How many table settings are there, in twenty-three boxes? — 138
10. How many centimetres are there, in twenty-four point one metres? — 2410 cm

11. A driver has completed fifty-six kilometres of a journey of length one hundred and two kilometres. How far has he yet to drive? — 46 km
12. Multiply seventeen by eight. — 136
13. To nought point one five, add nought point one five. — 0.3
14. Look at your answer sheet. Estimate the size of the angle in degrees. — (40°)
15. A dentist can see seven people in a session. How many sessions will be needed for the dentist to see thirty people? — 5
16. What is fifty less than four hundred and forty-four? — 394
17. Multiply two hundred and twelve by six. — 1272
18. Ten per cent of a seven hundred and fifty millilitre bottle of liquid has been used. What quantity of liquid has been used? — 75 ml
19. Write one and a quarter hours, in minutes. — 75 min
20. Three quarters of a candle, of original length sixty millimetres has been used. What length of candle is left? — 15 mm

21. Add these numbers: seven hundred and sixty-seven, and five hundred and eighty-five. — 1352
22. How many fourteens are there, in four hundred and twenty? — 30
23. How much would remain of a one hundred-pound budget, after the purchase of three suitcases costing twenty-six pounds each? — £22
24. What is three thousand, eight hundred, divided by eight? — 475
25. Multiply forty-four by eleven. — 484

C19

Answers

1 Divide two thousand, four hundred and ten by one hundred.   24.1
2 What is two squared?   4
3 In a hall, there are forty-three rows of chairs, and ten chairs in each row. How many chairs are there, altogether?   430
4 What are seven threes?   21
5 Write down a multiple of seven, which is less than twenty?   7 or 14
6 How many millimetres are there, in eight and a half centimetres?   85 mm
7 From five, subtract eight.   –3
8 Seventy people have each paid a travel agent three hundred pounds for a cruise holiday. What is the total amount of money taken by the travel agent?   £21000
9 Multiply forty-two by seven.   294
10 How many kilometres are there, in seven thousand metres?   7 km

11 The water was turned off at quarter past four. It came back on at five past five. For how long was the water turned off?   50 min
12 To five point nought three, add two point four.   7.43
13 What is twice thirty-seven?   74
14 Multiply fifty-seven by three.   171
15 How many halves are there, in five and a half whole ones?   11
16 Twenty-five per cent of a four hundred kilogram load of sand has been removed from a lorry. What is the weight of sand removed?   100 kg
17 Write down all the prime numbers between twenty and twenty-five.   23
18 Multiply four hundred and fifteen by four.   1660
19 A gallery has two hundred and eighty-three paintings in storage, but another thirty-eight paintings arrive. How many paintings will there be, altogether?   321
20 Write one hundred and fifty minutes, in hours. $2\frac{1}{2}$ hrs

21 Add four hundred and thirty-five to seven hundred and ninety-eight.   1233
22 How many sixes are there, in five thousand, two hundred and two.   867
23 What number is halfway between six point five and eight point five?   7.5
24 Divide three hundred and forty by seventeen.   20
25 Look at the map on your answer sheet, which has a scale of one centimetre to ten kilometres. Estimate the direct distance from point A to point B, in kilometres.   (37 km)

C20

Answers

1 A box is sold containing six cakes. How many boxes can be packed with six cakes each, using a batch of forty-eight cakes?   8
2 Multiply eighty-one by one hundred.   8100
3 How many tens are there, in two hundred and seventy?   27
4 How many fifths are there, in three whole ones?   15
5 What are nine fours?   36
6 How many metres are there, in two hundred and eighty centimetres?   2.8 m
7 How many grams are there, in nine point four five kilograms?   9450 g
8 Multiply twenty by nine hundred.   18000
9 Write in figures: three quarters of a million.   750000
10 To minus four, add minus two.   –6

11 An interview is expected to last for thirty-five minutes. It starts at ten to five. At what time is it expected to end?   5.25*
12 What is twice forty-six?   92
13 To three point two, add one point nought four.   4.24
14 What is sixty-five, multiplied by four?   260
15 Twenty-five pupils were absent, out of a year group of two hundred and nine. How many pupils were present?   184
16 Write three and two thirds of an hour, in minutes.   220 min
17 What is four fifths of sixty grams?   48 g
18 What is forty-three more than five hundred and seventy-eight?   621
19 Look at the line AB on your answer sheet. Write down an estimate for the length of the line, in centimetres.   (5 cm)
20 One eighth of a number is two point five. What is the number?   20

21 How many fours are there, in three thousand, eight hundred and twenty-eight?   957
22 Multiply fifteen by eleven.   165
23 A driver leaves home at quarter to nine, and returns at twenty past five that evening. How long, in hours and minutes, has the driver been away from home?   8 hrs 35 min
24 How many eighteens are there, in five hundred and forty?   30
25 Ronald sells an antique for eight hundred and eighty-seven pounds, and another for three hundred and forty-five pounds. What is the total amount of his sales?   £1232

51

1 A shop has three cages, and there are eight rabbits in each cage.
How many rabbits does the shop have?   24

2 What is five squared?   25

3 How many tens are there, in four hundred and ten?   41

4 How many pence are there, in forty-six pounds?   4600 p

5 How many rows of nine chairs each can be set out with sixty-three chairs?   7

6 What is seven thousand, multiplied by eight hundred?   5600000

7 From seven, subtract nine.   −2

8 How many centimetres are there, in seventy-three millimetres?   7.3 cm

9 Approximately how many kilograms are there, in twenty-two pounds weight?   10 kg

10 How many millilitres are there, in two point one two litres?   2120 ml

11 A television programme is on from twenty to six to twenty-five past six.
How long is the programme?   45 min

12 To one point three, add two point six.   3.9

13 What is forty-five, multiplied by six?   270

14 What is the greatest number of twenty-pence coins you could get from seven pounds?   35

15 Look at your answer sheet.
Estimate the size of the angles, in degrees.   (50°)

16 Fifty per cent of a two hundred and fifty millilitre medicine bottle has been used.
What quantity of medicine has been used?   125 ml

17 From two hundred and three, subtract forty-eight.   155

18 Multiply one hundred and twenty-three by four.   492

19 What is three eighths of one hundred and twenty?   45

20 Write two and a third hours, in minutes.   140 min

21 Add these numbers: eight hundred and forty-six, and four hundred and seventy-seven.   1323

22 How many sixteens are there, in three hundred and twenty?   20

23 An aeroplane flight lasts from fourteen forty hours to seventeen twenty hours.
How long was the flight, in hours and minutes.   2 hrs 40 min

24 What is four thousand, one hundred and eighty-five, divided by nine?   465

25 Multiply twenty-two by twelve.   264

1 A woman drives eight kilometres each day to and from work.
How far will she drive, over a period of nine days?   72 km

2 How many ninths are there, in five whole ones?   45

3 What is six hundred and ten, multiplied by ten?   6100

4 What is five hundred and forty-three, divided by one hundred?   5.43

5 What is three times three times three?   27

6 How many centimetres are there, in forty-three millimetres?   4.3 cm

7 Multiply six thousand by two hundred.   1200000

8 A temperature of minus three degrees rises by seven degrees.
What is the new temperature?   4°

9 How many litres are there, in seven hundred millilitres?   0.7 litres

10 A car park has room for sixteen cars in each of three areas.
What is the maximum number of cars that can be parked?   48

11 Eight per cent of pupils are late for school on a particular day.
What percentage are on time?   92%

12 What is forty-one, multiplied by eight?   328

13 What is seven minus five point two?   1.8

14 What is twenty-eight less than one hundred and five?   77

15 A box contains six eggs.
How many boxes can be filled from a batch of fifty eggs?   8

16 Multiply one hundred and twenty-four by three.   372

17 Write two and a third hours, in minutes.   140 min

18 Ten per cent of a number is fifty.
What is the number?   500

19 Sally has six hundred and nine pounds, but spends fifty-seven pounds.
How much has she left?   £552

20 On your answer sheet, draw an arrow on the protractor scale to indicate your estimate of a thirty-degree angle.   (30°)

21 Prize money of two thousand, nine hundred and ten pounds is to be divided equally between six people.
How much will each person receive?   £485

22 What is fifty-three, multiplied by eleven?   583

23 How many twelves are there, in eight hundred and forty?   70

24 How much change do you get from twenty pounds, if you buy three ties at four pounds and fifty pence each?   £6.50

25 Add eight hundred and seventy-eight, and three hundred and forty-five.   1223

C23

1. Seven photographs can fit on the page of a photo album.
   How many pages will be needed for forty-two photographs? — 6
2. Divide one hundred and thirty-nine by ten. — 13.9
3. What are nine fours? — 36
4. Multiply twenty-four by one hundred. — 2400
5. How many six-seater rowing boats will be needed by forty-eight women? — 8
6. How many metres are there, in six hundred and twenty-four centimetres? — 6.24 m
7. A CD costs nine pounds.
   What will be the total cost of twenty-one CDs, at the same price each? — £189
8. Multiply seventy by thirty. — 2100
9. How many grams are there, in fifteen point two kilograms? — 15200 g
10. A temperature of four degrees falls by six degrees. What is the new temperature? — −2°

11. Twenty-five per cent of a 600-metre path is overgrown.
    What length of the path is overgrown? — 150 m
12. What is forty-seven less than one hundred and one? — 54
13. How many days are there, in sixteen weeks? — 112
14. What is a quarter, written as a percentage? — 25%
15. Look at your answer sheet.
    Write down an estimated reading from the dial. — (5.78 g)
16. To two point one, add three point five. — 5.6
17. What is one fifth of ninety pounds? — £18
18. Multiply three hundred and twenty-three by five. — 1615
19. Add two hundred and eighty-seven, and seventy-six. — 363
20. Write one hundred and thirty-five minutes, in hours. — $2\frac{1}{4}$ hrs

21. What number is halfway between seven point five and fifteen point five? — 11.5
22. Three hundred and sixty degrees is to be divided into twelve equal sectors.
    How many degrees will there be in each sector? — 30°
23. Add eight hundred and thirty-seven to five hundred and eighty-four. — 1421
24. What is six thousand and fifty-five, divided by seven? — 865
25. Look at the map on your answer sheet, which has a scale of one centimetre to five kilometres.
    Estimate the direct distance from point A to point B, in kilometres. — (14 km)

C24

1. What is the total cost of nine packs of cards, if each pack costs three pounds? — £27
2. What is six squared? — 36
3. Multiply forty-three by ten. — 430
4. How many sevenths are there, in two whole ones? — 14
5. Divide six hundred and ninety by one hundred. — 6.9
6. A temperature falls from five degrees to minus two degrees.
   By how many degrees has the temperature fallen? — 7°
7. Multiply four hundred by nine hundred. — 360000
8. How many kilometres are there, in four thousand, five hundred metres? — 4.5 km
9. A ferry crosses a 77-kilometre stretch of water four times in one day.
   What is the total distance it had travelled? — 308 km
10. Approximately how many pounds weight are there, in two kilograms? — 4.4 lb

11. Seventy-six per cent of pupils in a room are boys.
    What percentage are girls? — 24%
12. To two point one six, add nought point nought four. — 2.2
13. What is twice fifty-three? — 106
14. What is seventy-three, multiplied by four? — 292
15. Sean spends forty-three pounds of his one hundred and nine pounds savings.
    How much will he have left? — £66
16. Look at your answer sheet.
    Estimate the length of the line AB, in millimetres. — (24 mm)
17. What is three hundred and forty-one, multiplied by eight? — 2728
18. What is two thirds of six thousand lire? — 4000 lire
19. What is fifty-eight less than four hundred and five? — 347
20. How many grams is ninety per cent of two hundred and fifty grams? — 225 g

21. How much change should you get from fifty pounds, if you buy three concert tickets at twelve pounds and fifty pence each? — £12.50
22. What number is halfway between four point two and six point two? — 5.2
23. How many seventeens are there, in three hundred and forty? — 20
24. There are eight hundred and four pupils in a hall.
    They are joined by the other three hundred and fifty-seven pupils from the rest of the school.
    How many pupils are there, in the school? — 1161
25. Divide three thousand, nine hundred and twelve by four. — 978

1 What is three thousand, five hundred and fifty divided by one hundred?   35.5

2 Multiply ninety-four by ten.   940

3 How many weeks are there, in forty-two days?   6 wks

4 There are four wheels on a car. How many wheels are there, altogether, on six cars?   24

5 Write down a multiple of three, which is less than ten.   3, 6 or 9

6 How many millimetres are there, in six point four centimetres?   64 mm

7 What is the total value of sixteen five-pound notes?   £80

8 A temperature rises by seven degrees, from minus four degrees. What is the new temperature?   3°

9 Multiply eighty by seven hundred.   56000

10 How many litres are there, in eight thousand, seven hundred and ninety millilitres?   8.79 litres

11 Write down the length of the perimeter of a square of side of length three centimetres.   12 cm

12 Multiply thirty-nine by three.   117

13 What is nine minus six point four?   2.6

14 Lisa has walked forty-eight miles of a charity walk of length one hundred and seventeen miles. How far has she yet to walk?   69 miles

15 Look at your answer sheet. Write down an estimated reading from the dial.   (20.8 km/hr)

16 Three eighths of forty people are female. How many females are there?   15

17 What is seventy-nine less than three hundred and three?   224

18 Work out five per cent of three thousand kilograms.   150 kg

19 What is four hundred and four, multiplied by nine?   3636

20 Write two and a quarter hours, in minutes.   135 min

21 To seven hundred and forty-seven, add five hundred and ninety-six.   1343

22 How many twelves are there, in six hundred?   50

23 You buy four books at five pounds and fifty pence each. What change should you receive from thirty pounds?   £8

24 Multiply twenty-two by thirteen.   286

25 Divide six thousand, two hundred and eight by eight.   776

1 Seven cards are needed to play a game. There are thirty-five cards in a box. How many people can play the game?   5

2 What are eight sixes?   48

3 What is nine squared?   81

4 There are one hundred counters in a box. A school orders eighty-six boxes. What is the total number of counters ordered?   8600

5 What is twenty-seven, divided by ten?   2.7

6 How many kilograms are there, in ten thousand grams?   10 kg

7 From eight, subtract nine.   −1

8 Write a quarter of a million, in figures.   250000

9 There are three hundred screws in a box. How many screws will there be, in twenty boxes?   6000

10 How many centimetres are there, in thirty metres?   3000 cm

11 Tom started cutting his lawn at quarter past two, and finished at five past three. For how long was he cutting the lawn?   50 min

12 To three point nought seven, add one point three.   4.37

13 One sixth of a number is nine. What is the number?   54

14 Look at your answer sheet. Estimate the size of the angle, in degrees.   (70°)

15 A felt tip pen costs twenty-five pence. How much would it cost to buy nine felt tip pens?   £2.25

16 What is forty-seven less than three hundred and nineteen?   272

17 Fifty per cent of a number is seventy. What is the number?   140

18 Multiply three hundred and two by eight.   2416

19 Write three hundred and thirty minutes in hours. $5\frac{1}{2}$ hrs

20 What is three tenths of seventy kilograms?   21 kg

21 What number is halfway between three point five and nine point five?   6.5

22 How many sevens are there, in four thousand and ninety-five?   585

23 You buy three CDs at ten pounds, forty pence each. What change should you receive from forty pounds?   £8.80

24 Divide six hundred and eighty by forty.   17

25 A three-seater settee costs eight hundred and fifty-six pounds. A two-seater settee costs five hundred and sixty-seven pounds. What is the cost of one of each settee?   £1423

C27

Answers

1 How many full boxes of six eggs each can be packed from fifty-four eggs? — 9

2 How many days are there, in nine weeks? — 63 days

3 Multiply twenty-eight by ten. — 280

4 Divide three hundred and twenty by one hundred. — 3.2

5 A tin contains exactly eight plums. How many tins can be packed from fifty-six plums? — 7

6 Approximately how many kilometres are there, in ten miles? — 16 km

7 A temperature rises from minus three degrees to six degrees. By how many degrees has the temperature risen? — 9°

8 Multiply ten thousand by sixty. — 600000

9 How many metres are there, in one hundred and forty centimetres? — 1.4 m

10 A new computer game costs fifty-one pounds. How much would it cost to buy six games at the same price? — £306

11 What is the greatest number of ten-pence coins you could get for fifteen pounds? — 150

12 From eight point one two, subtract three point nought one. — 5.11

13 Multiply thirty-seven by nine. — 333

14 A woman slept for eighteen per cent of the day. For what percentage of the day was she awake? — 82%

15 How many fifths are there, in twelve whole ones? — 60

16 What is fifty-nine more than five hundred and seventy-two. — 631

17 Write three and a third hours, in minutes. — 200 min

18 Anne has complete two fifths of a fifty-five mile journey. How many miles has she gone? — 22 miles

19 Write down all the prime numbers between twenty-five and thirty. — 29

20 Multiply two hundred and forty-two by six. — 1452

21 What is three hundred and forty-seven, added to five hundred and ninety-eight? — 945

22 What is seven hundred, divided by fourteen? — 50

23 Prize money of five thousand, one hundred and ninety-three pounds is to be divided equally between nine people. How much will each person receive? — £577

24 What number is halfway between five point two and seven point two? — 6.2

25 Look at the map on your answer sheet, which has a scale of one centimetre to four kilometres. Estimate the direct distance from point A to point B, in kilometres. — (9.2 km)

C28

Answers

1 A minibus can carry nine passengers. How many minibuses will be needed for a group of forty-five passengers? — 5

2 What is one hundred and forty, divided by ten? — 14

3 How many eighths are there, in three whole ones? — 24

4 What are six sevens? — 42

5 How many pence are there, in sixty-two pounds? — 6200 p

6 Helen needs seven small bikes for her playschool. They cost fifty-six pounds each. What will be the total cost? — £392

7 How many centimetres are there, in sixty-nine millimetres? — 6.9 cm

8 Multiply four hundred by sixty. — 24000

9 How many grams are there, in seventy-four point five kilograms? — 74500 g

10 What is minus two, added to minus five? — −7

11 One hundred and fourteen parents attend a meeting. Thirty-nine are male. How many are female? — 75

12 What is three point nought one, add one point three three? — 4.34

13 A plane began its descent at eighteen thirty-five hours, and landed at nineteen twenty-five hours. How long did it take to descend and land? — 50 min

14 What is twenty-three, multiplied by eight? — 184

15 One of a hundred people, twenty-two travelled regularly on a train. Write this as a percentage. — 22%

16 Find twenty per cent of three hundred and fifty. — 70

17 What is one third of ninety-six kilograms? — 32 kg

18 On your answer sheet, draw an arrow on the protractor scale to indicate your estimate of a forty-five degree angle. — (45°)

19 What is three hundred and five, multiplied by seven? — 2135

20 A cabinet priced at three hundred and fifty pounds is reduced by seventy-four pounds. What is its new price? — £276

21 A car journey is expected to take three hours and ten minutes. If the journey starts at twenty past eleven, at what time should it finish? — 14.30*

22 How many nineteens are there, in five hundred and seventy? — 30

23 Multiply forty-three by twelve. — 516

24 What is four thousand and sixty-eight, divided by six? — 678

25 To five hundred and sixty-two, add six hundred and eighty-nine. — 1251

C29

1. What is the total cost of nine video tapes, at three pounds each? — £27
2. Multiply thirty-five by one hundred. — 3500
3. What is two times two times two? — 8
4. How many tens are there, in four hundred and twenty? — 42
5. Write down a multiple of six, which is less than twenty. — 6, 12 or 18
6. How many litres are there, in three thousand, two hundred millilitres? — 3.2 litres
7. From four, subtract seven. — −3
8. A box for posting weighs nine kilograms. What is the total weight of forty-one boxes of the same weight? — 369 kg
9. How many metres are there, in sixteen kilometres? — 16000 m
10. Multiply seventy by thirty. — 2100

11. What is the greatest number of ten-pence coins you can get for nineteen pounds? — 190
12. What is twice eighty-six? — 172
13. Multiply twenty-eight by four. — 112
14. One seventh of a number is eight. What is the number? — 56
15. From five, subtract three point one. — 1.9
16. A man is given a discount of sixty per cent on an insurance policy which costs one hundred and fifty pounds. How much is the discount? — £90
17. What is five eighths of twenty-four hours? — 15 hrs
18. Multiply one hundred and fifty-four by four. — 616
19. On the line AB on your answer sheet, put an X at a point which you estimate to be three centimetres from point A. — (3 cm)
20. A booking for a car ferry costs three hundred and eighty-nine pounds for a car, and eighty-nine pounds for a trailer. What is the total cost of booking a car with a trailer? — £478

21. What is four hundred and eighty-seven, added to six hundred and thirty-seven? — 1124
22. How many fourteens are there, in four hundred and twenty? — 30
23. You buy four train tickets costing five pounds, fifty pence each. What change should you receive from thirty pounds? — £8
24. What number is halfway between seven point five and fifteen point five? — 11.5
25. Divide three thousand, nine hundred and fifty-two by four. — 988

C30
Answers

1. What is the result of dividing three thousand, five hundred by one hundred? — 35
2. How many quarters are there, in two whole ones? — 8
3. What is nine squared? — 81
4. What is thirty, multiplied by ten? — 300
5. How many nine centimetres lengths can be cut from a 72-centimetre length of wood? — 8
6. How many centimetres are there, in one and a half metres? — 150 cm
7. Approximately, how many pints are there, in two litres? — $3\frac{1}{2}$ pints
8. There are three hundred metal washers in a box. How many metal washers will there be, in seventy similar boxes? — 21000
9. From four, subtract seven. — −3
10. How many litres are there, in four thousand and eighty millilitres? — 4.08 litres

11. From nine point three three, subtract three point nought three. — 6.3
12. What is fifty-five less than one hundred and four? — 49
13. A video machine taped a fifty-five minute programme starting at twenty thirty hours. At what time did it finish taping? — 21.25*
14. What is the total cost of three car tyres which cost seventy-seven pounds each? — £231
15. Look at your answer sheet. Write down an estimated reading from the dial. — 4.27 mm
16. One quarter of a sixty-eight gram block of cheese has been eaten. What is the weight of cheese that has been eaten? — 17 g
17. What is three hundred and two, multiplied by nine? — 2718
18. Write down all the prime numbers between thirty and thirty-five. — 31
19. Find fifteen per cent of four thousand. — 600
20. Write one and two thirds of an hour, in minutes. — 100 min

21. The day in a school starts at ten past nine, and finishes at twenty to four. How long, in hours and minutes, is the school day? — 6 hrs 30 min
22. Add six hundred and sixty-four to five hundred and forty-eight. — 1212
23. How many elevens are there, in four hundred and forty? — 40
24. What number is halfway between three point seven and seven point seven? — 5.7
25. Divide six thousand, seven hundred and fourteen by nine. — 746

1. A holiday lasts for fifty-six days. How many weeks are there, in fifty-six days? — 8 wks
2. Write down a factor of six. — 1, 2, 3 or 6
3. What are four threes? — 12
4. What is eight hundred and fifty, divided by one hundred? — 8.5
5. Multiply thirty-three by ten. — 330
6. A temperature falls by eight degrees, from six degrees. What is the new temperature? — −2°
7. Three pipes of length eighty-seven centimetres are fixed together, end to end. What will be the total length? — 261 cm
8. Multiply two hundred by thirty. — 6000
9. How many kilograms are there, in sixteen thousand grams? — 16 kg
10. How many metres are there, in three thousand, two hundred centimetres? — 32 m

11. Write down the length of the perimeter of a square of side of length four centimetres. — 16 cm
12. To seven point five one, add one point one one. — 8.62
13. Multiply forty-three by seven. — 301
14. A box contains eight pens. How many full boxes can be packed from a supply of sixty pens? — 7
15. What is eighty-five less than one hundred and twenty-three? — 38
16. What is two hundred and thirty-four, multiplied by three? — 702
17. Write two hundred and ten minutes, in hours. — $3\frac{1}{2}$ hrs
18. A computer is reduced by seventy-five pounds from five hundred and three pounds. What is the new price? — £428
19. What is forty per cent of one hundred and eighty? — 72
20. Find seven eighths of twenty-four. — 21

21. What number is halfway between three point five and nine point five? — 6.5
22. One dozen strawberries is twelve strawberries. How many dozen strawberries are there, in seven hundred and twenty? — 60
23. What is five hundred and forty-seven, added to six hundred and ninety-seven? — 1244
24. What is six thousand, one hundred and thirty-nine, divided by seven? — 877
25. Look at the map on your answer sheet, which has a scale of one centimetre to ten kilometres. Estimate the direct distance from point A to point B, in kilometres. — (28 km)

1. A pig sty can hold a maximum of eight pigs. How many sties will be needed for a herd of seventy-two pigs? — 9
2. How many fifths are there, in three whole ones? — 15
3. What is four hundred and twenty, divided by ten? — 42
4. What is eighty-five, multiplied by one hundred? — 8500
5. What are seven sixes? — 42
6. What is the total cost of four wheels, when one wheel costs fifty-three pounds? — £212
7. How many centimetres are there, in eighty-one millimetres? — 8.1 cm
8. From minus three, subtract minus two. — −1
9. Multiply seven hundred by seventy. — 49000
10. How many kilometres are there, in three thousand, two hundred and fifty metres? — 3.25 km

11. A table tennis match started at five to four, and finished at twenty to five. How long was the game? — 45 min
12. From five point seven eight, subtract three point seven two. — 2.06
13. Multiply thirty-two by eight. — 256
14. A double-decker bus can carry one hundred and ten passengers, but a single-decker can only carry forty-eight. How many more passengers can be carried by the double-decker bus? — 62
15. Look at your answer sheet. Estimate the size of the angle, in degrees. — (30°)
16. What is three tenths of four hundred? — 120
17. What is three hundred and twelve, multiplied by four? — 1248
18. Twenty-five per cent of a number is one hundred. What is the number? — 400
19. To five hundred and eighty-seven, add sixty-eight. — 655
20. Write two and one third hours, in minutes. — 140 min

21. What is four thousand, four hundred and eighty-eight, divided by six? — 748
22. Add five hundred and eighty-six to six hundred and fifty-four. — 1240
23. Brian has spent four and a quarter hours on a ferry. He arrives at his destination at five past ten. At what time did he depart? — 5.50*
24. How many elevens are there, in seven hundred and seventy? — 70
25. What number is halfway between four point five and fourteen point five? — 9.5

1 How many nine-centimetre lengths can be cut from a 54-centimetre length of wood? — 6
2 What is five squared? — 25
3 What is fifty-six, divided by ten? — 5.6
4 What is twenty-four, multiplied by one hundred? — 2400
5 There are three points to a triangle. What is the total number of points on nine separate triangles? — 27
6 How many litres are there, in five thousand, two hundred and thirty millilitres? — 5.23 litres
7 Multiply seven thousand by thirty. — 210000
8 Write in figures: three quarters of a million. — 750000
9 A temperature rises from minus two degrees to six degrees. By how many degrees has it risen? — 8°
10 There are forty-five teams in a five-a-side tournament. How many players is this, altogether? — 225

11 On a certain day, eight per cent of pupils were absent from school. What percentage was present? — 92%
12 How many sixths are there, in twelve whole ones? — 72
13 Multiply seventy-two by nine. — 648
14 From eight, subtract six point three. — 1.7
15 What is thirty-six less than one hundred and twenty? — 84
16 Four fifths of fifty-five people questioned, disliked going to work. How many people disliked going to work? — 44
17 Multiply four hundred and one by eight. — 3208
18 Of two hundred and three blood donors, fifty-four have had a cold and cannot give blood. How many are able to give blood? — 149
19 What is fifty per cent of five hundred and fifty? — 275
20 On the line AB on your answer sheet, put an X at a point which you estimate to be four centimetres from point A. — (4 cm)

21 A power cut lasts from five past eleven to ten to three. How long, in hours and minutes, was the power cut? — 3 hrs 45 min
22 Multiply sixteen by twelve. — 192
23 How many elevens are there, in five hundred and fifty? — 50
24 Divide five thousand, three hundred and twelve by eight. — 664
25 On the outward journey, a ferry carried seven hundred and seventy-five cars. On the return, the number was four hundred and sixty-eight. How many cars were carried, altogether? — 1243

1 An octagon has eight points. How many points are there, altogether, on six octagons? — 48
2 What is three times three times three? — 27
3 What is three hundred and thirty, divided by one hundred? — 3.3
4 Write down a multiple of seven, which is less than twenty. — 7 or 14
5 Four out of every ten people on a bus are male. Write this as a fraction. — $\frac{4}{10}$ or $\frac{2}{5}$
6 How many centimetres are there, in six hundred and sixty millimetres? — 66 cm
7 From six, subtract nine. — −3
8 Multiply seven hundred by nine hundred. — 630000
9 How many kilometres are there, in seven thousand metres? — 7 km
10 Approximately how many feet are there, in a length of ninety centimetres? — 3 ft

11 Alan has a forty-minute bike ride. He left home at twenty past ten. At what time should he return? — 11.00*
12 To two point three seven, add five point nought two. — 7.39
13 What is seventy-four, multiplied by six? — 444
14 A television costing one hundred and five pounds has its price reduced by thirty-eight pounds in a closing-down sale. What is the sale price? — £67
15 Look at your answer sheet. Estimate the size of the angle, in degrees. — 40°
16 What is ninety per cent of three hundred and thirty? — 297
17 Six boxes weigh three hundred and fourteen grams each. What is their total weight? — 1884 g
18 From six hundred and nine, subtract seventy-six. — 533
19 What is two thirds of thirty-nine? — 26
20 Write three and three quarter hours, in minutes. — 225 min

21 A gardener took four hours and twenty minutes to dig the borders of his garden. He finished at three o'clock. At what time did he start? — 10.40*
22 How many thirties are there, in five hundred and forty? — 18
23 Multiply twelve by fifteen. — 180
24 Add three hundred and fifty-six to eight hundred and sixty-eight. — 1224
25 Divide three thousand, five hundred and twelve by four. — 878

C35

1. An eight-a-side tournament is being set up.
   How many teams can be made up from forty-eight players? — 6
2. Multiply forty-six by ten. — 460
3. What are seven nines? — 63
4. What is five hundred and thirty-two, divided by one hundred? — 5.32
5. Three out of every ten people in a class do not like chips.
   Write this as a fraction. — $\frac{3}{10}$
6. What is four thousand, multiplied by forty? — 160000
7. How many kilograms are there, in ninety thousand grams. — 90 kg
8. A temperature of minus five degrees rises by four degrees.
   What is the new temperature? — –1°
9. A protractor costs six pence.
   How much would it cost, in pence, to buy protractors for a class of twenty-eight children? — 168 p
10. How many centimetres are there, in four metres? — 400 cm

11. A coach can carry sixty-seven people.
    How many people can be carried by three of these coaches? — 201
12. What is thirty-four less than one hundred and fourteen. — 80
13. One sixth of a number is eight.
    What is the number? — 48
14. From seven point five four, subtract two point four four. — 5.1
15. What is the greatest number of twenty-pence coins you could get for nine pounds? — 45
16. Write down all the prime numbers between thirty-five and forty. — 37
17. What is four hundred and eleven, multiplied by nine? — 3699
18. On your answer sheet, draw an arrow on the protractor scale to indicate your estimate of a sixty-degree angle. — (60°)
19. A train has covered three fifths of a three hundred and fifty mile journey.
    How far has it travelled, in miles? — 210 miles
20. Add forty-seven to six hundred and ninety-two. — 739

21. How many weeks are there, in four thousand, six hundred and six days? — 658 wks
22. Multiply forty-four by eleven. — 484
23. How many fifteens are there, in nine hundred. — 60
24. Add six hundred and fifty-five to seven hundred and eighty-six. — 1441
25. Look at the map on your answer sheet, which has a scale of one centimetre to five kilometres.
    Estimate the direct distance from point A to point B, in kilometres. — (15 km)

C36

1. A holiday lasts for forty-nine days.
   How many weeks are there, in forty-nine days? — 7 wks
2. How many tens are there, in three hundred and fifty? — 35
3. Multiply fifty-six by one hundred. — 5600
4. How many ninths are there, in two whole ones? — 18
5. A machine can make sixty-two tools in a day.
   How many can be made over a seven-day period? — 434
6. What are four eights? — 32
7. Subtract nine from five. — –4
8. How many metres are there, in forty-eight point two kilometres. — 48200 m
9. Multiply seven hundred by forty. — 28000
10. How many litres are there, in seven thousand millilitres? — 7 litres

11. In one week, fifty-six per cent of people passed their driving test.
    What percentage failed? — 44%
12. From seven, subtract five point four. — 1.6
13. Multiply thirty-four by seven. — 238
14. A tank with a maximum capacity of one hundred and eight litres has forty-five litres of oil inside.
    How much more oil can be added to the tank? — 63 litres
15. Look at your answer sheet.
    Write down an estimated reading from the dial. — (0.8 litres)
16. Seventy-five per cent of a four thousand pound grant has been spent.
    How much money has been spent? — £3000
17. Multiply three hundred and sixteen by five. — 1580
18. What is one eighth of one hundred and twenty? — 15
19. Write one and two thirds of an hour, in minutes. — 100 min
20. A dining-room table costing eight hundred and four pounds is reduced in price by sixty-five pounds.
    What is its new price? — £739

21. What number is halfway between seven point four and nine point four? — 8.4
22. How many thirteens are there, in two hundred and sixty? — 20
23. An Intercity train leaves Glasgow at eleven forty hours, and arrives at London Euston at sixteen fifty hours.
    How long, in hours and minutes, was the journey? — 5 hrs 10 min
24. Add three hundred and fifty-four to eight hundred and seventy-eight. — 1232
25. Divide six thousand and twelve by nine. — 668

1   There are nine sausage rolls in a bag.
    How many sausage rolls would you
    get in six bags?   54
2   Multiply fifty-seven by one hundred.   5700
3   Write down a multiple of two, which
    is less than ten.   2, 4, 6 or 8
4   What is four squared?   16
5   Divide nine hundred and forty-seven
    by ten.   94.7
6   How many millimetres are there, in
    twelve point eight centimetres?   128 mm
7   Multiply six hundred by six hundred.   360000
8   A temperature of four degrees falls by
    seven degrees.
    What is the new temperature?   –3°
9   How many kilometres are there, in six
    thousand, five hundred metres?   6.5 km
10 A delivery man travels eight times a
    week between two towns which are
    sixty-six kilometres apart.
    What is the total distance travelled
    each week?   528 km

11 A radio programme started at
    eighteen thirty hours, and finished at
    nineteen ten hours.
    How long was the programme?   40 min
12 What is twice fifty-four?   108
13 To three point three three, add six point
    nought five.   9.38
14 Multiply sixty-three by four.   252
15 A taxi can carry six people.
    How many taxis are needed for forty
    people?   7
16 What is eighty-seven less than six
    hundred and twenty-one?   534
17 Multiply three hundred and sixteen
    by three.   948
18 Look at your answer sheet.
    Estimate the length of the line AB,
    in millimetres.   (30 mm)
19 What is one quarter of ninety-six
    gallons?   24 gal
20 A man pays income tax of twenty per
    cent on his earnings of four hundred
    and fifty pounds.
    How much income tax does he pay?   £90

21 Add these numbers: five hundred
    and seven, and six hundred and
    eighty-four.   1191
22 How many nineteens are there, in three
    hundred and eighty?   20
23 Multiply thirty-one by twelve.   372
24 How much change should you get from
    twenty-five pounds, if you buy three
    calculators costing six pounds and fifty
    pence each?   £5.50
25 Divide six thousand, two hundred and
    eighty by eight.   785

1   A company can make one hundred
    watches in a week.
    How many weeks will it take to fill
    an order for two thousand, five
    hundred watches?   25 wks
2   What are seven eights?   56
3   How many quarters are there, in five
    whole ones?   20
4   Write down a factor of fifteen.   1, 3, 5 or 15
5   How many hundreds are there, in
    two thousand, five hundred.   25
6   How many kilograms are there, in
    sixteen thousand grams?   16 kg
7   Multiply four thousand by eight
    hundred.   3200000
8   A temperature of five degrees falls by
    seven degrees.
    What is the new temperature?   –2°
9   There are forty-three chocolates in a
    box. How many chocolates will you get
    in nine boxes?   387
10 Approximately, how many kilograms are
    there, in six point six pounds weight?   3 kg

11 James is thirty-six centimetres smaller
    than his brother Paul.
    Paul is one hundred and twenty
    centimetres tall.
    How tall is James?   84 cm
12 Multiply fifty-four by six.   324
13 From three point five one, subtract
    one point one one.   2.4
14 Look at your answer sheet.
    Write down an estimated reading
    from the dial.   (8.2 kg)
15 A lesson lasted from twenty-five past
    nine to ten past ten.
    How long was the lesson?   45 min
16 From five hundred and three, subtract
    fifty-four.   449
17 Ten per cent of a number is seven
    hundred. What is the number?   7000
18 What is four hundred and twenty-one,
    multiplied by seven?   2947
19 There are ninety tickets available for a
    play. Three tenths have been sold.
    How many tickets have been sold?   27
20 Write three and two thirds of an hour,
    in minutes.   220 min

21 Add these numbers: nine hundred
    and four to three hundred and
    fifty-seven.   1261
22 How many thirteens are there, in six
    hundred and fifty.   50
23 Multiply fifty-seven by eleven.   627
24 Shreena works from half past eleven
    until quarter past four.
    For how long, in hours and minutes,
    does she work?   4 hrs 45 min
25 Divide five thousand, two hundred and
    sixty-eight by six.   878

1 What is the result of dividing two thousand, five hundred by one hundred?   25
2 What is seven squared?   49
3 What are nine sixes?   54
4 On a bus, seven out of every ten people are making a short journey. Write this as a fraction. $\frac{7}{10}$
5 Multiply fourteen by ten.   140
6 How many metres are there, in seven hundred and forty centimetres?   7.4 m
7 There are sixty minutes in an hour. How many minutes are there, in ninety hours?   5400 min
8 Multiply seventy-eight by three.   234
9 From four, subtract eight.   −4
10 How many millilitres are there, in seven point six litres?   7600 ml

11 A suit costing one hundred and forty pounds is reduced by twenty-seven pounds. What is the new price of the suit?   £113
12 Multiply thirty-five by nine.   315
13 What is a half, written as a percentage?   50%
14 To three point nought two, add two point two two.   5.24
15 Look at your answer sheet. Estimate the size of the angle, in degrees.   (80°)
16 Multiply four hundred and three by five.   2015
17 A mountaineer has climbed thirty per cent of a cliff of height three hundred and seventy feet. How high has the mountaineer climbed, in feet?   111 ft
18 What is sixty-three more than seven hundred and ninety-eight?   861
19 Find seven eighths of twenty-four.   21
20 What is one hundred minutes, written in hours? $1\frac{2}{3}$ hrs

21 Add these numbers: nine hundred and ninety-six to two hundred and thirty-nine.   1235
22 How many sixteens are there, in eight hundred?   50
23 You buy four train tickets costing seven pounds and fifty pence each. What change should you receive from fifty pounds?   £20
24 Divide four thousand, eight hundred and nine by seven.   687
25 What number is halfway between seven point four and eleven point four?   9.4

1 A century means one hundred years. How many years are there, in seventy-five centuries?   7500 yrs
2 What are six sevens?   42
3 How many tens are there, in nine hundred?   90
4 Write down a multiple of four, which is less than fifteen.   4, 8 or 12
5 What is two times two times two?   8
6 There are fifty-two sweets in a packet. How many sweets are there, in four packets?   208
7 What is seven thousand, multiplied by sixty?   420000
8 To minus three, add minus two.   −5
9 How many kilometres are there, in seven thousand, eight hundred and twenty metres?   7.82 km
10 Write in figures: half a million.   500000

11 Write down the length of the perimeter of a square of side of length five centimetres.   20 cm
12 What is eighty-five less than one hundred and twenty-three?   38
13 From three point nought seven, subtract one point nought two.   2.05
14 Multiply eighteen by eight.   144
15 There was a power cut from twenty past six to five past seven. How many minutes was this?   45 min
16 What is four hundred and twenty-seven, multiplied by three?   1281
17 What is half of one hundred and fifty metres?   75 m
18 On your answer sheet, draw an arrow on the protractor scale to indicate your estimate of a seventy-degree angle.   (70°)
19 From three hundred and three, subtract ninety-eight.   205
20 Gerry is given a discount of five per cent off an order of value eight hundred pounds. How many pounds are knocked off the order?   £40

21 Divide six thousand, two hundred and twenty-four by eight.   778
22 How many elevens are there, in three hundred and thirty?   30
23 Two items of jewellery cost four hundred and eighty-five pounds, and eight hundred and forty-seven pounds. What is the total cost of the jewellery?   £1332
24 What number is halfway between three point seven and seven point seven?   5.7
25 Look at the map on your answer sheet, which has a scale of one centimetre to five kilometres. Estimate the direct distance from point A to point B, in kilometres.   12.5 km

1 Approximately how many centimetres are there, in a two-foot length?   60 cm
2 How many grams are there, in four point two kilograms?   4200 g
3 Subtract nine from five.   −4
4 How many hundreds are there, in two thousand, five hundred?   25
5 Multiply forty by twenty.   800
6 Write two fifths, as a decimal.   0.4
7 Multiply five point one eight by one hundred.   518
8 Write seventy per cent, as a fraction in its simplest form. $\frac{7}{10}$
9 What is minus three, multiplied by minus four?   12
10 Write twenty-one hundredths, as a percentage.   21%

11 A trip lasted one and three quarters of an hour.
How many minutes is this?   105 min
12 Find three fifths of thirty grams.   18 g
13 Twenty-five per cent of a number is twenty. What is the number?   80
14 What is eighty-four less than two hundred and three?   119
15 On your answer sheet, draw an arrow on the protractor scale to indicate your estimate of a forty-five degrees angle.   (45°)
16 The three angles of a quadrilateral are eighty degrees, ninety degrees, and one hundred degrees.
What is the size of the fourth angle?   90°
17 Look at the equation on your answer sheet: y is twice x, taken away from thirteen.
If x equals five, what is y?   3
18 Increase five hundred and ten grams by ten per cent.   561 g
19 One eighth of a number is two point five. What is the number?   20
20 Decrease thirty by one fifth.   24

21 Look at the map on your answer sheet, which has a scale of one centimetre to five kilometres.
Estimate the direct distance from point A to point B, in kilometres.   (15 km)
22 A bingo prize of four hundred and eighty pounds is divided equally between sixteen people.
How much does each person receive?   £30
23 What is six squared plus two squared?   40
24 Look at the equation on your answer sheet.
What is x? $3\frac{1}{2}$
25 What is fifty-nine, multiplied by eleven?   649

1 How many millimetres are there, in nought point seven centimetres?   7 mm
2 How many centimetres are there, in five and a half metres?   550 cm
3 What is seven point four five, multiplied by one thousand?   7450
4 From two, subtract ten.   −8
5 What is one hundred, multiplied by twenty?   2000
6 Write fifty-three hundredths, as a decimal.   0.53
7 What is the area of a rectangle of length eight centimetres and width six centimetres?   48 cm²
8 What is thirteen point three, divided by one thousand?   0.0133
9 Write one tenth, as a percentage.   10%
10 Write thirty-five per cent, as a fraction in its lowest terms. $\frac{7}{20}$

11 Malcolm has four hundred and ninety-one pounds.
He earns another forty-eight pounds.
How much will he have now?   £539
12 What is three fifths of ninety kilograms?   54 kg
13 Write down all the prime numbers between ten and fifteen.   11, 13
14 Write two and two thirds of an hour, in minutes.   160 min
15 What is one hundred and twenty-three, multiplied by four?   492
16 The perimeter of a square is forty centimetres.
What is its area?   100 cm²
17 Write, as a decimal, three divided by twenty-five.   0.12
18 Look at the equation on your answer sheet: C is twice the sum of d and three.
If d is eight, what is C?   22
19 A ferry is carrying five hundred car passengers and two hundred and ninety foot passengers. How many passengers are there, altogether?   790
20 Reduce sixty-four pounds by twenty-five per cent.   £48

21 Add these numbers: two hundred and forty-seven to three hundred and ninety-eight.   645
22 How many twelves are there, in three hundred and sixty?   30
23 What is sixteen, multiplied by twelve?   192
24 The mean of five numbers is twenty-one.
What is the sum of the five numbers?   105
25 From seven squared, subtract three squared.   40

D3

Answers

1. How many kilometres are there, in two thousand, four hundred metres? — 2.4 m
2. Multiply seven point four by one hundred. — 740
3. A temperature of eight degrees falls by ten degrees. What is the new temperature? — −2°
4. How many millilitres are there, in three and a half litres? — 3500 ml
5. Multiply eight hundred by six hundred. — 480000
6. Write seven hundredths, as a percentage. — 7%
7. Divide three point nought seven by one hundred. — 0.0307
8. Write nought point four one, as a percentage. — 41%
9. What is minus six, divided by minus three? — 2
10. Write one quarter, as a decimal. — 0.25

11. Clare makes seven trips each of two hundred and forty-one kilometres in length. What is her total mileage for the trips? — 1687 km
12. What is half of nineteen centimetres? — 9.5 cm
13. What is ten per cent of two hundred and fifty pounds? — £25
14. What is sixty-four less than three hundred and two? — 238
15. On the line AB on your answer sheet, put an X at a point which you estimate to be three centimetres from point A. — (3 cm)
16. Two angles of a triangle are eighty degrees and sixty degrees. What size is the third angle? — 40°
17. The ratio of cars to lorries passing a point on a road in one hour is nine to two. There were fourteen lorries. How many cars were there? — 63
18. Multiply four hundred and forty by forty. — 17600
19. Increase two hundred and twenty litres by seventy per cent. — 374 litres
20. Look at the equation on your answer sheet: p is the square of the sum of q and two. Find p, when q is seven. — 81

21. What number is halfway between five point seven and fifteen point seven? — 10.7
22. How many fourteens are there, in two hundred and eighty? — 20
23. Divide eight hundred and seventy-six by six. — 146
24. Find the mean of three, five, nine and eleven. — 7
25. Look at the equation on your answer sheet. Find y, when x is five. — 16

D4

Answers

1. How many kilometres are there, in nine thousand, five hundred metres? — 9.5 km
2. How many centimetres are there, in eight and a half metres? — 850 cm
3. Divide one hundred and eighty-three by ten. — 18.3
4. To minus three, add five. — 2
5. Multiply seven thousand by twenty. — 140000
6. Write sixty-three hundredths, as a decimal. — 0.63
7. What is nine point eight one, multiplied by one thousand? — 9810
8. Write seven tenths, as a percentage. — 70%
9. Find the area of a rectangle of length seven metres and width four metres. — 28 m²
10. Write ninety-three per cent, as a decimal. — 0.93

11. Sally has one hundred and seven teddy bears. She sells thirty-eight. How many teddy bears has she left? — 69
12. Approximately how many inches are there, in two metres? — 80 in
13. Multiply five hundred and twelve by four. — 2048
14. Seventy-five per cent of one thousand pounds raised by a school is donated to charity. How much money is given to charity? — £750
15. What is one eighth of forty-eight litres? — 6 litres
16. Three of the angles of a quadrilateral are one hundred and ten degrees, one hundred and twenty degrees, and eighty degrees. What is the size of the fourth angle? — 50°
17. One third of a number is six. What is the number? — 18
18. A path thirty-eight metres long is increased in length by fifty per cent. What is the new length? — 57 m
19. Look at the equation on your answer sheet: a is two times c squared. If c is nine, what is a? — 162
20. Decrease thirty by one sixth. — 25

21. Look at the map on your answer sheet, which has a scale of one centimetre to four kilometres. Estimate the direct distance from point A to point B, in kilometres. — (10 km)
22. What are thirteen twelves? — 156
23. How many thirties are there, in five hundred and forty? — 18
24. Take two squared from three squared. — 5
25. Look at the equation on your answer sheet. What is x? — 3

D5 — Answers

1. How many metres are there, in four hundred and eighty centimetres? — 4.8 m
2. How many seconds are there, in two hundred minutes? — 1200 sec
3. Write half a million, in figures. — 500000
4. How many kilometres are there, in three thousand, seven hundred and fifty metres? — 3.75 km
5. From two, subtract four. — −2
6. Write three tenths, as a decimal. — 0.3
7. What is eighty-seven point nine, multiplied by one thousand? — 87900
8. Write fifteen per cent, as a fraction in its lowest terms. — $\frac{3}{20}$
9. Write thirty-seven hundredths, as a percentage. — 37%
10. Find the area of a triangle of height six metres and base of length four metres. — 12 m²

11. A coat costing fifty-six pounds is reduced by one quarter in a sale. What is the sale price of the coat? — £42
12. Add fifty-nine to three hundred and ninety-four. — 453
13. What is three quarters of sixty? — 45
14. Multiply three hundred and one by six. — 1806
15. On your answer sheet, draw an arrow on the protractor scale to indicate your estimate of a thirty-degree angle. — (30°)
16. Multiply two hundred and thirty by three hundred. — 69000
17. Twenty per cent of a number is fifty. What is the number? — 250
18. The perimeter of a rectangle is thirty-four centimetres. The width is seven centimetres. What is its area? — 70 cm²
19. Write, as a percentage, seventeen divided by twenty-five. — 68%
20. Look at the equation on your answer sheet: y is x squared, minus four. If x equals thirteen, what is y? — 165

21. A new settee costs eight hundred and fifty-seven pounds. A chair costs four hundred and seventy-five pounds. What is the total cost of one settee and one chair? — £1332
22. Multiply seventeen by fifteen. — 255
23. How many fifteens are there, in nine hundred? — 60
24. From nine squared, take six squared. — 45
25. Find the mean of five, one, one, three, and five. — 3

D6 — Answers

1. How many kilograms are there, in six thousand, five hundred grams? — 6.5 kg
2. To minus six, add plus four. — −2
3. Multiply eight hundred by fifty. — 40000
4. Approximately how many litres are there, in two gallons? — 9 litres
5. How many millilitres are there, in five point three litres? — 5300 ml
6. Write four fifths, as a decimal. — 0.8
7. What is minus eight divide by minus two? — 4
8. Write eighty-one hundredths, as a percentage. — 81%
9. What is nineteen point one, multiplied by one hundred? — 1910
10. Write thirty-seven per cent, as a decimal. — 0.37

11. Five people are each paid two hundred and three pounds. What is the total amount earned? — £1015
12. What is three fifths of fifty-five? — 33
13. Look at your answer sheet. Estimate the size of the angle, in degrees. — (70°)
14. What is thirty per cent of one hundred and twenty milligrams? — 36 mg
15. Write one and a quarter hours, in minutes. — 75 min
16. What is forty, divided by four thousand? Give your answer as a decimal. — 0.01
17. Thirty-five grams are divided into two parts in the ratio of three to two. What is the heavier weight? — 21 g
18. Two of the angles of a triangle are seventy degrees and sixty-five degrees. What is the size of the third angle? — 45°
19. Increase twenty-one by one seventh. — 24
20. Look at the equation on your answer sheet: a is twenty-four, divided by the sum of c and three. Find a, when c is three. — 4

21. The mean of four numbers is eight. What is the sum of the four numbers? — 32
22. How many elevens are there, in two hundred and twenty? — 20
23. Look at the equation on your answer sheet. What is y, if x is seven? — 12
24. Divide five hundred and forty-nine by nine. — 61
25. What are fourteen twelves? — 168

1 There are fourteen places on each row in church.
What is the greatest number of people who can sit in the first three rows? 42

2 What is five thousand, multiplied by three hundred? 1500000

3 A temperature of minus two degrees rises by seven degrees. What is the new temperature? 5°

4 How many litres are there, in eight hundred millilitres? 0.8 litres

5 How many centimetres are there, in forty-two millimetres? 4.2 cm

6 Write one quarter, as a percentage. 25%

7 What is nought point nought two two, multiplied by one thousand? 22

8 What is the area of a rectangle of length eight metres and width seven metres? 56 m²

9 Write fifty-nine hundredths, as a decimal. 0.59

10 Write nought point eight five, as a fraction in its lowest terms. $\frac{17}{20}$

11 There are seven hundred and nine people on a ferry.
Sixty-seven people get off the ferry. How many people remain? 642

12 Multiply one hundred and twenty-six by three. 378

13 Write two and one third hours, in minutes. 140 min

14 What is seven tenths of fifty grams? 35 g

15 Write down all the prime numbers between fifteen and twenty. 17, 19

16 Multiply four hundred and eighty by four hundred. 192000

17 Mortgage payments of ninety-two pounds are reduced by one per cent. What is the new payment? £91.08

18 Look at the equation on your answer sheet: *d* is four times the answer to *e* subtract two.
Find *d*, when *e* is ten. 32

19 The perimeter of a square is sixteen centimetres.
What is its area? 16 cm²

20 What is eight increased by a half of eight? 12

21 Six hundred pupils are divided into fifteen groups.
How many pupils are there, in each group? 40

22 What number is halfway between three point five and eleven point five? 7.5

23 Add four hundred and eighty-five to seven hundred and forty-six. 1231

24 Find the mean of two, four, two, seven, and five. 4

25 Add eight squared to four squared. 80

1 Three friends are celebrating their joint twenty-first birthdays.
What is their total age? 63 yrs

2 How many metres are there, in seven hundred and twenty-four centimetres? 7.24 m

3 A temperature of three degrees falls by six degrees.
What is the new temperature? –3°

4 Multiply sixty by forty. 2400

5 How many kilograms are there, in fifty-one point two grams? 0.0512 kg

6 Write thirteen thousandths, as a decimal. 0.013

7 What is seventy-eight point five, divided by one thousand? 0.0785

8 Write three fifths, as a percentage. 60%

9 What is minus three, multiplied by minus four? 12

10 Write thirty-five per cent, as a fraction in its lowest terms. $\frac{7}{20}$

11 An 800-metre pipe is reduced in length by twenty-five per cent.
By what length is it reduced? 200 m

12 Add seventy-eight to one hundred and eighty-seven. 265

13 What is one fifth of ninety pounds? £18

14 How much money is two hundred and thirty-two five-pound notes? £1160

15 On your answer sheet, draw an arrow on the protractor scale to indicate your estimate of a sixty-degree angle. (60°)

16 Write, as a decimal, nine divided by fifty. 0.18

17 Three angles of a quadrilateral are eighty degrees, one hundred and ten degrees, and ninety degrees.
What is the size of the fourth angle? 80°

18 Look at the equation on your answer sheet: *y* is six times *x* squared.
Find *y*, when *x* is seven. 294

19 Multiply five hundred and twenty by seventy. 36400

20 A measure of one hundred and twenty millilitres of fluid in a bottle is increased by thirty per cent. How much is then in the bottle? 156 ml

21 Look at the map on your answer sheet, which has a scale of one centimetre to ten kilometres.
Estimate the direct distance from point A to point B, in kilometres. (35 km)

22 Multiply twenty-eight by eleven. 308

23 Divide two hundred and sixty by thirteen. 20

24 From five squared, subtract three squared. 16

25 Look at the equation on your answer sheet. What is *x*? 7

1 How many kilometres are there, in five thousand, five hundred metres? — 5.5 km

2 What is seven hundred, multiplied by five hundred? — 350000

3 A temperature of minus two degrees falls by a further four degrees. What is the new temperature? — −6°

4 Multiply eighty-seven by four. — 348

5 Approximately how many pounds weight are there, in two kilograms? — 4.4 lb

6 What is the area of a triangle with a height of five metres, and length of base eight metres? — 20 m²

7 Write nine tenths, as a decimal. — 0.9

8 What is four point nine nine, multiplied by one thousand? — 4990

9 Write forty-three hundredths, as a percentage. — 43%

10 Write three quarters of a million, in figures. — 750000

---

11 Two angles in a triangle are sixty degrees and ninety degrees. What is the size of the third angle? — 30°

12 From four hundred and five, subtract sixty-eight. — 337

13 Look at your answer sheet. Estimate the length of the line AB, in centimetres. — (2 cm)

14 What is ninety per cent of three hundred and fifty grams? — 315 g

15 Multiply four hundred and thirty-one by eight. — 3448

16 What is two thirds of nine thousand lire? — 6000 lire

17 Reduce three hundred and thirty by sixty per cent. — 132

18 Multiply three hundred and ten by three hundred. — 93000

19 Look at the equation on your answer sheet: d is the square of the sum of e and three. Find d, when e is four. — 49

20 Two lengths are in the ratio of three to seven. The second length is forty-two centimetres. Find the first length. — 18 cm

---

21 What number is halfway between seven point five and nine point five? — 8.5

22 Divide three hundred and eighty by nineteen. — 20

23 Nine hundred and seventy-six cards are divided into four piles. How many cards are there, in each pile? — 244

24 The mean of six numbers is ten. What is the sum of the six numbers? — 60

25 Look at the equation on your answer sheet. What is y, when x is seven? — 38

1 There are five days in a working week. How many days are there, in fourteen working weeks? — 70 days

2 How many millimetres are there, in five point four centimetres. — 54 mm

3 Multiply four hundred by ninety. — 36000

4 A temperature of minus four degrees rises to six degrees. By how many degrees has the temperature risen? — 10°

5 How many litres are there, in seven thousand, eight hundred and ninety millilitres? — 7.89 litres

6 Write nine hundredths, as a decimal. — 0.09

7 What is nought point three nine two, multiplied by one hundred? — 39.2

8 What is minus two, multiplied by six? — −12

9 Write one fifth, as a percentage. — 20%

10 Write nought point four seven, as a fraction. — $\frac{47}{100}$

---

11 Three eighths of forty people in a hall are male. How many are male? — 15

12 There are eight hundred and three pupils in a school. Sixty-nine children do not go on a school trip. How many do go on the trip? — 734

13 What is five per cent of four thousand? — 200

14 Multiply five hundred and four by nine. — 4536

15 Write two and one quarter of an hour, in minutes. — 135 min

16 The perimeter of a rectangle is ten metres. The length of the rectangle is three metres. Find its area. — 6 m²

17 Write, as a percentage, nine divided by twenty-five. — 36%

18 What is sixty, divided by six hundred? — 0.1

19 Look at the equation on your answer sheet: y is x squared, plus three. Find y, when x is ten. — 103

20 Reduce thirty-six by one sixth. — 30

---

21 There are twelve pens in a box. What is the total number of pens in sixteen boxes? — 192

22 How many sixteens are there, in three hundred and twenty? — 20

23 Look at the map on your answer sheet, which has a scale of one centimetre to twenty kilometres. Estimate the direct distance from point A to point B, in kilometres. — (74 km)

24 Subtract two squared from four squared. — 12

25 Look at the equation on your answer sheet. What is x? — 5

D11

1 What is the area of a rectangle with length nine metres, and width seven metres? 63 m²

2 Multiply three hundred by twenty. 6000

3 How many kilograms are there, in ten thousand grams? 10 kg

4 Write one quarter of a million, in figures. 250000

5 How many centimetres are there, in twenty metres? 2000 cm

6 Write thirty-seven hundredths, as a percentage. 37%

7 From eight, subtract nine. −1

8 What is seven point four six, divided by one hundred? 0.0746

9 Write three quarters, as a decimal. 0.75

10 Write nought point four five, as a percentage. 45%

11 Martin reduced a track record of two hundred and nineteen seconds by thirty-seven seconds.
What is the new record? 182 sec

12 What is nine tenths of seventy kilograms? 63 kg

13 What is two hundred and three, multiplied by eight? 1624

14 Fifty per cent of a number is sixty. What is the number? 120

15 On the line AB on your answer sheet, put an X at a point which you estimate to be thirty millimetres from point A. (30 mm)

16 Increase ten pounds by forty per cent. £14

17 Look at the equation on your answer sheet: C is twenty-five, minus three x. Find C, when x is seven. 4

18 A mass of twenty grams is reduced by two fifths.
What is the new mass? 12 g

19 The perimeter of a rectangle is ten centimetres, and its length is three centimetres.
What is its area? 6 cm²

20 Multiply four thousand, six hundred by two hundred. 920000

21 Which number is exactly halfway between six point five and ten point five? 8.5

22 A van can travel a distance of thirteen miles using one gallon of petrol.
How many gallons are needed for a 650-mile journey? 50 gal

23 Add eight hundred and forty-seven to three hundred and seventy-eight. 1225

24 The mean of three numbers is sixteen. What is the sum of the three numbers? 48

25 From eight squared, take four squared. 48

D12

1 Approximately, how many kilometres are equivalent to ten miles? 16 km

2 Multiply ten thousand by fifty. 500000

3 A temperature of minus three degrees rises by five degrees.
What is the new temperature? 2°

4 How many metres are there, in one hundred and twenty centimetres? 1.2 m

5 There are six eggs in a box.
What is the total number of eggs in forty-one boxes? 246

6 Write twenty-seven hundredths, as a decimal. 0.27

7 What is fifty-one point six, divided by one thousand? 0.0516

8 What is the area of a triangle of height six metres, and length of base eight metres? 24 m²

9 Write nine tenths, as a percentage. 90%

10 Write eighty-one per cent, as a fraction. $\frac{81}{100}$

11 Three angles of a quadrilateral are one hundred degrees, one hundred and thirty degrees, and seventy degrees.
What is the size of the fourth angle of the quadrilateral? 60°

12 Write three and one third hours, in minutes. 200 min

13 Add fifty-nine to four hundred and ninety-two. 551

14 What is two fifths of fifty-five miles? 22 miles

15 What is one hundred and forty-two, multiplied by six? 852

16 Multiply two thousand, seven hundred by forty. 108000

17 One fifth of a number is three point five. What is the number? 17.5

18 Look at the equation on your answer sheet: d is sixteen, divided by the sum of c and three.
Find d, when c is one. 4

19 A sum of two hundred pounds is divided between two people in the ratio of three to two.
What is the smallest share? £80

20 Increase twelve by three quarters. 21

21 One week is seven days.
How many weeks are there, in eight hundred and sixty-one days. 123 wks

22 How many twenty-fives are there, in six hundred and fifty? 26

23 Multiply fifteen by fifteen. 225

24 Find the mean of two, five, three, eight, nine, and three. 5

25 Look at the equation on your answer sheet. What is y, when x is 6? 28

1 A child is seventy-six weeks old.
How many days are there, in seventy-six
weeks?   532 days

2 How many centimetres are there, in
sixty-nine millimetres?   6.9 cm

3 Multiply four hundred by fifty.   20000

4 How many grams are there, in
eighty-four point five kilograms?   84500 g

5 To minus two, add minus three.   –5

6 Write ninety-seven thousandths, as a
decimal.   0.097

7 What is sixty-three point six, divided
by one hundred?   0.636

8 What is plus ten, divided by minus two?   –5

9 Write nought point nine five, as a
percentage.   95%

10 Write four fifths, as a percentage.   80%

---

11 Barry's share of a bonus is one third of
ninety-six pounds.
How much is Barry's share?   £32

12 What is seventy-four less than two
hundred and two?   128

13 What is twenty per cent of four hundred
and fifty kilograms?   90 kg

14 What is two hundred and five,
multiplied by seven?   1435

15 On your answer sheet, draw an arrow
on the protractor scale to indicate your
estimate of an eighty-degree angle.   (80°)

16 Write, as a decimal, eleven divided by
twenty-five.   0.44

17 Look at the equation on your answer
sheet: p is the square of q plus three.
Find p, when q is four.   49

18 What is ninety, divided by nine
thousand?
Give your answer as a decimal.   0.01

19 The perimeter of a square is eight
centimetres.
What is the area of the square?   4 cm²

20 The usual water level in a reservoir is
forty metres.
During a drought, the water levels falls
by eighty per cent.
During the drought, by how many
metres does the water level fall?   32 m

---

21 Eight hundred pupils are divided
equally between sixteen coaches for a
school trip.
How many pupils are on each coach?   50

22 What are forty-six elevens?   506

23 Add eight hundred and eighty-six to
four hundred and seventy-five.   1361

24 Subtract three squared from four squared. 7

25 Look at the equation on your answer
sheet. What is x?   8

1 How many litres are there, in two
thousand, two hundred and ten
millilitres?   2.21 litres

2 Multiply thirty-one by nine.   279

3 How many metres are there, in
fourteen kilometres?   14000 m

4 From four, subtract seven.   –3

5 What is eighty, multiplied by twenty?   1600

6 Write one fifth, as a decimal.   0.2

7 What is four point four one, multiplied
by one hundred?   441

8 What is the area of a square of sides of
length seven centimetres?   49 cm²

9 Write eight hundredths, as a percentage. 8%

10 Write nought point four five, as a
fraction in its lowest terms. $\frac{9}{20}$

---

11 Ali spends three eighths of the
twenty-four hours of the day fast asleep.
For how long does he sleep?   9 hrs

12 Find sixty per cent of one hundred and
seventy pounds.   £102

13 Multiply two hundred and forty-five by
four.   980

14 Add thirty-nine to four hundred and
eighty-nine.   528

15 On the line AB on your answer sheet, put
an X at a point which you estimate to
be twenty millimetres from point A.   (20 mm)

16 The two angles of a triangle are
eighty-five degrees and forty-five degrees.
What is the size of the third angle in
the triangle?   50°

17 Multiply seven thousand, one hundred
by five hundred.   3550000

18 A worker's weekly wage of one hundred
pounds is increased by nine per cent.
What will be the new weekly wage?   £109

19 Look at the equation on your answer
sheet: y is three times, x minus two.
Find y, when x is twelve.   30

20 Reduce twelve grams by one quarter.   9 g

---

21 There are eight coasters to a box.
How many boxes can be filled from
six hundred and eighty-seven coasters?   85

22 What is four hundred and twenty,
divided by fourteen?   30

23 What are twenty-two elevens?   242

24 What is eight squared, added to two
squared?   68

25 Find the mean of eight, twelve, eight,
and sixteen.   11

D15

1 A ribbon is one and a half metres long. How many centimetres are there, in one and a half metres? **150 cm**
2 Multiply two hundred by sixty. **12000**
3 How many litres are there, in six thousand and eighty millilitres? **6.08 litres**
4 From four, subtract eight. **−4**
5 Approximately how many pints are there, in two litres? **3$\frac{1}{2}$ pints**
6 Write sixty-nine hundredths, as a decimal. **0.69**
7 What is thirteen point nought one, divided by one hundred? **0.1301**
8 Write three quarters, as a percentage. **75%**
9 What is the area of a triangle of height nine metres, and base of length six metres? **27 m^2**
10 Write eighty-three per cent, as a fraction. **$\frac{83}{100}$**

11 On your answer sheet, write down all the prime numbers between twenty and twenty-five. **23**
12 Write one and a third hours, in minutes. **80 min**
13 What is one quarter of sixty-eight pounds? **£17**
14 Multiply four hundred and two by nine. **3618**
15 What is eighty-eight less than two hundred and seven? **119**
16 The area of a rectangle is twelve square metres. The width of the rectangle is three metres. What is the perimeter? **14 m**
17 Increase fourteen by two sevenths. **18**
18 Look at the equation on your answer sheet: d is equal to three times, e squared. Find d, when e is nine. **243**
19 Multiply two hundred and eight by twenty. **4160**
20 The ratio of boys to girls in a class is four to three. There are twenty-four girls in the class. How many boys are there? **32**

21 The mean of five numbers is thirty-two. What is the sum of the numbers? **160**
22 What is the mean of three point five and five point five? **4.5**
23 Look at the map on your answer sheet, which has a scale of one centimetre to five kilometres. Estimate the direct distance from point A to point B, in kilometres. **(12.5 km)**
24 Divide three hundred and forty by seventeen. **20**
25 Look at the equation on your answer sheet. Find y, when x is nine. **56**

D16

1 A temperature of six degrees falls by eight degrees. What is the new temperature? **−2°**
2 Multiply three hundred by twenty. **6000**
3 How many kilograms are there, in fifteen thousand grams? **15 kg**
4 Multiply seventy-eight by three. **234**
5 How many metres are there, in two thousand, three hundred centimetres? **23 m**
6 What is nought point nought three eight, multiplied by one thousand? **38**
7 What is minus four, multiplied by three? **−12**
8 What is seven tenths, as a decimal? **0.7**
9 What is sixty-five per cent, as a decimal? **0.65**
10 What is thirty-nine hundredths, as a percentage? **39%**

11 Brenda has driven seven eighths of her 24-kilometre journey. How many kilometres has she driven? **21 km**
12 What is three hundred and twenty-four, multiplied by three? **972**
13 What is forty per cent of one hundred and eighty milligrams? **72 mg**
14 There are six hundred and three seats for a play in the hall. Seventy-five seats are vacant. What is the attendance for the play? **528**
15 On the line AB on your answer sheet, put an X at a point which you estimate to be four centimetres from point A. **(4 cm)**
16 The perimeter of a square is twenty-four millimetres. What is its area? **36 mm^2**
17 Multiply one hundred and ninety by twenty. **3800**
18 Look at the equation on your answer sheet: a equals twelve, divided by the sum of c and two. Find a, when c is six. **1.5**
19 Decrease two pounds by nine per cent. **£1.82**
20 One sixth of a number is twelve. What is the number? **72**

21 What is eight hundred and seventy-six, added to six hundred and ninety-eight? **1574**
22 Divide three hundred and thirty by eleven. **30**
23 One dozen is twelve. How many is twenty-four dozen? **288**
24 To five squared, add two squared. **29**
25 Look at the equation on your answer sheet. What is x? **7**

1 How many kilometres are there, in four thousand, two hundred and fifty metres? — 4.25 km

2 From minus three, subtract minus two. — –1

3 Multiply eight hundred by eighty. — 64000

4 How many centimetres are there, in ninety-one millimetres? — 9.1 cm

5 Multiply forty-three by four. — 172

6 Write twenty-three hundredths, as a percentage. — 23%

7 What is eighty-three point one, multiplied by one hundred? — 8310

8 What is the area of a rectangle of length nine metres and width four metres? — 36 m²

9 Write nought point nine, as a percentage. — 90%

10 Write one fifth, as a decimal. — 0.2

---

11 A bill for fifteen pounds is increased by one hundred per cent. What is the amount of the revised bill? — £30

12 What is three tenths of three hundred millimetres? — 90 mm

13 To four hundred and eighty-seven, add sixty-eight. — 555

14 Multiply two hundred and thirteen by four. — 852

15 On your answer sheet, draw an arrow on the protractor scale to indicate your estimate of a forty-five degree angle. — (45°)

16 Ten per cent of a number is thirty. What is the number? — 300

17 Multiply five thousand, one hundred by eighty. — 408000

18 Look at the equation on your answer sheet: y equals x squared, plus four. Find y, when x is four. — 20

19 The three angles of a quadrilateral are eighty degrees, ninety degrees, and eighty degrees. What is the size of the fourth angle? — 110°

20 Write, as a decimal, thirteen divided by twenty-five. — 0.52

---

21 Eight hundred and sixty-eight leaflets are to be distributed by seven volunteers. If they are divided equally, how many will each volunteer have to deliver? — 124

22 How many sixteens are there, in eight hundred? — 50

23 What are seventy-six elevens? — 836

24 Find the mean of three, nine, five, and three. — 5

25 What is five squared less than nine squared? — 56

1 How many litres are there, in six thousand, two hundred and ten millilitres? — 6.21 litres

2 Write in figures: three quarters of a million. — 750000

3 Multiply three thousand by seventy. — 210000

4 A temperature of minus two degrees rises to five degrees. By how many degrees has the temperature risen? — 7°

5 How much is thirty-five notes, all of which are five-pound notes? — £175

6 What is nine, divided by minus three? — –3

7 Write eleven hundredths, as a decimal. — 0.11

8 Write two fifths, as a percentage. — 40%

9 What is three point three seven, divided by one hundred? — 0.0337

10 Write fifty-five per cent, as a fraction in its lowest terms. — $\frac{11}{20}$

---

11 A new car radio is reduced by fifty-four pounds from two hundred and three pounds. What is the new price? — £149

12 What is fifty per cent of five hundred and fifty millilitres. — 275 ml

13 What is six hundred and one, multiplied by eight? — 4808

14 What is four fifths of forty-five people. — 36

15 On the line AB on your answer sheet, put an X at a point which you estimate to be three centimetres from point A. — (3 cm)

16 Multiply seven hundred and twenty by forty. — 28800

17 Reduce eighty grams by five per cent. — 76 g

18 The perimeter of a rectangle is twenty-four centimetres. The length of the rectangle is seven centimetres. What is its area? — 35 cm²

19 Look at the equation on your answer sheet: p equals twenty, minus four q. Find p, when q is five. — 0

20 Eighty-four is divided in the ratio of four to three. What is the smaller share? — 36

---

21 Which number is exactly halfway between five point five and nine point five? — 7.5

22 How many eighteens are there, in five hundred and forty? — 30

23 To eight hundred and forty-six, add four hundred and sixty-seven. — 1313

24 The mean of five numbers is thirty-two. What is the sum of the five numbers? — 160

25 Look at the equation on your answer sheet. What is y, when x is seven? — 44

D19

1. How many centimetres are there, in seven hundred and sixty millimetres? — 76 cm
2. From seven, subtract nine. — −2
3. How many kilometres are there, in eight thousand metres. — 8 km
4. What is nine hundred, multiplied by seven hundred? — 630000
5. Approximately how many feet are there, in a length of ninety centimetres? — 3 ft
6. What is nought point seven five five, multiplied by one thousand? — 755
7. Write seven hundredths, as a decimal. — 0.07
8. What is the area of a triangle of height four metres, and base of length eight metres? — 16 m²
9. Write three tenths, as a percentage. — 30%
10. Write fourteen per cent, as a decimal. — 0.14

11. A machine can produce two hundred and fourteen glasses in an hour. How many can it produce in six hours? — 1284
12. What is ninety per cent of three hundred and thirty? — 297
13. What is seventy-six less than five hundred and nine? — 433
14. Find two thirds of thirty-nine dollars. — $26
15. Write three and a quarter hours, in minutes. — 195 min
16. One quarter of a number is twenty. What is the number? — 80
17. Three angles of a quadrilateral are one hundred degrees, eighty degrees, and one hundred degrees. What is the size of the fourth angle? — 80°
18. Multiply two hundred and fifty by seventy. — 17500
19. A weight of seventy-seven kilograms is increased by two hundred per cent. What is the new weight? — 231 kg
20. Look at the equation on your answer sheet: y equals twenty, divided by the sum of x and two. Find y, when x is two. — 5

21. Which number is halfway between four point six and six point six? — 5.6
22. Look at the map on your sheet, which has a scale of one centimetre to ten kilometres. Estimate the direct distance from point A to point B, in kilometres. — (22.5 km)
23. What is three squared, added to two squared? — 13
24. How many sixteens are there, in three hundred and twenty? — 20
25. Find the mean of five, seven, two, and ten. — 6

D20

1. A temperate of minus five degrees rises by three degrees. What is the new temperature? — −2°
2. How many kilograms are there, in fifty thousand grams? — 50 kg
3. Multiply two thousand by forty. — 80000
4. How many centimetres are there, in five metres? — 500 cm
5. Multiply twenty-seven by six. — 162
6. Write thirty-seven hundredths, as a percentage. — 37%
7. What is seventy-nine point six, divided by one hundred? — 0.796
8. Write three fifths, as a decimal. — 0.6
9. What is minus eight, divided by minus two? — 4
10. Write ninety-three per cent, as a fraction. — $\frac{93}{100}$

11. A seven hundred and ninety-two pound bill for central heating installation is increased by forty-seven pounds. What is the total bill? — £839
12. On your answer sheet, draw an arrow on the protractor scale to indicate your estimate of a sixty-degree angle. — (60°)
13. What is five hundred and eleven, multiplied by nine? — 4599
14. What is fifty-five per cent of six thousand? — 3300
15. What is three fifths of four hundred and fifty pounds? — £270
16. Multiply five thousand, six hundred by thirty. — 168000
17. Two sevenths of a class of twenty-one are absent. How many children are present? — 15
18. Look at the equation on your answer sheet: d equals double the answer to e plus three. Find d, when e is fifteen. — 36
19. Increase eighty by ten per cent. — 88
20. The perimeter of a square is four centimetres. What is its area? — 1 cm²

21. What is the total cost, in pence, of eleven stamps, each of value fifteen pence. — 165p
22. How many seventeens are there, in eight hundred and fifty? — 50
23. What is eight hundred and ninety-six, divided by four? — 224
24. What is eight squared, added to five squared? — 89
25. Look at the equation on your answer sheet. What is x? — 4

1 Seven people have each received an eighty-two pound share of prize money won on the football pools.
   What was the total amount of prize money?   £574
2 From five, subtract eight.   –3
3 How many litres are there, in eight thousand millilitres?   8 litres
4 What is four hundred, multiplied by seventy?   28000
5 How many metres are there, in sixty-eight point one kilometres?   68100 m
6 Write one tenth, as a decimal.   0.1
7 What is three point seven two, multiplied by one hundred?   372
8 Write eighty-three hundredths, as a percentage.   83%
9 What is twenty, divided by minus five?   –4
10 Write nought point seven, as a fraction. $\frac{7}{10}$

11 Seventy-five per cent of an 8000-litre tank of diesel has been used.
   What quantity of diesel has been used?   6000 litres
12 Multiply two hundred and sixteen by five.   1080
13 What is sixty-five less than nine hundred and four?   839
14 Write one and two thirds hours, in minutes.   100 min
15 What is one eighth of one hundred and twenty kilometres?   15 km
16 In a rectangle, the ratio of the width to the length is five to eight.
   The width is twenty-five centimetres.
   Find the length.   40 cm
17 Look at the equation on your answer sheet: y equals five x squared.
   Find y, when x is seven.   245
18 Write, as a percentage, eleven divided by fifty.   22%
19 What is ninety, divided by one thousand, eight hundred?   0.05
20 The perimeter of a rectangle is sixteen centimetres.
   The width of the rectangle is two centimetres.
   What is its area?   12 cm²

21 A box contains eighteen corn forks.
   How many boxes can be made up from seven hundred and twenty corn forks?   40
22 One dozen is twelve.
   How many is seventeen dozen?   204
23 Add eight hundred and ninety-seven to four hundred and fifty-four.   1351
24 The mean of six numbers is thirteen.
   What is the sum of the six numbers?   78
25 Look at the equation on your answer sheet. What is y, when x is seven?   41

1 How many kilometres are there, in seven thousand, five hundred metres?   7.5 km
2 A temperature of minus four degrees falls by three degrees.
   What is the new temperature?   –7°
3 Multiply five hundred by five hundred.   250000
4 On average, one chicken on Mr McDonald's farm will lay eight eggs a week.
   How many eggs will be laid, on average, in one week by fifty-six chickens?   448
5 How many millimetres are there, in eleven point seven centimetres?   117 mm
6 What is nine point eight nine, divided by one hundred?   0.0989
7 Write three tenths, as a percentage.   30%
8 Write nine hundredths, as a decimal.   0.09
9 What is the area of a rectangle of length six metres, and width four metres? 24 m²
10 Write nought point five six, as a percentage.   56%

11 Look at your answer sheet.
   Estimate the length of the line AB, in centimetres.   (5 cm)
12 Aisha pays twenty per cent tax on her earnings.
   How much tax will she pay on four hundred and fifty pounds?   £90
13 Subtract eighty-seven from seven hundred and twenty-one.   634
14 What is one quarter of ninety-six litres?   24 litres
15 Multiply four hundred and sixteen by three.   1248
16 Write, as a decimal, three divided by fifty.   0.06
17 Two angles of a triangles are seventy degrees, and eighty degrees.
   What is the size of the third angle in the triangle?   30°
18 Look at the equation on your answer sheet: a equals c squared, plus five.
   Find a, when c is three.   14
19 What is six thousand, four hundred, multiplied by thirty?   192000
20 A jacket costing forty-four pounds is reduced by twenty-five per cent in a sale.
   What is the sale price of the jacket?   £33

21 Which number is halfway between six point five and fourteen point five?   10.5
22 What is twelve, divided into eight hundred and forty?   70
23 How many nines are there, in six hundred and eighty-four?   76
24 Find the mean of three, six, five, seven, four, and five.   5
25 What is two squared less than nine squared?   77

1 Approximately how many kilograms are there, in six point six pounds weight? **3 kg**

2 How many kilograms are there, in eighteen thousand grams? **18 kg**

3 There are twenty-three children in a class. If they each need nine counters for an investigation, what is the total number of counters needed by the class? **207**

4 A temperature of five degrees falls by another eight degrees. What is the new temperature? **−3°**

5 Multiply three thousand by eight hundred. **2400000**

6 Write nought point four five, as a fraction in its lowest terms. $\frac{9}{20}$

7 Write three quarters, as a percentage. **75%**

8 Multiply nought point nine six by one thousand. **960**

9 Write fifty-one hundredths, as a decimal. **0.51**

10 Multiply four by minus four. **−16**

---

11 Three fifths of the seventy tickets for a school play have been sold. How many tickets have been sold? **42**

12 What is four hundred and three, subtract sixty-four? **339**

13 Twenty per cent of a number is eighty. What is the number? **400**

14 What is three hundred and eleven, multiplied by seven? **2177**

15 Write three and two thirds hours, in minutes. **220 min**

16 Reduce thirty millilitres by four fifths. **6 ml**

17 What is the total value of seven hundred and forty televisions, each costing two hundred pounds? **£148000**

18 Look at the equation on your answer sheet: a equals six c squared. Find a, when c is four. **96**

19 The perimeter of a square is thirty-two metres. What is its area? **64 m²**

20 A machine can make forty combs every minute. Its output is increased by a fifth. How many combs can now be produced each minute? **48**

---

21 Look at the map on your sheet, which has a scale of one centimetre to five kilometres. Estimate the direct distance from point A to point B, in kilometres. **(15 km)**

22 How many twelves are there, in three hundred and sixty? **30**

23 What is the mean of seven point five and eleven point five? **9.5**

24 What is six squared plus five squared? **61**

25 Look at the equation on your answer sheet. What is x? **9**

1 How many metres are there, in eight hundred and forty centimetres? **8.4 m**

2 Multiply sixty-eight by three. **204**

3 How many millilitres are there, in eight point six litres? **8600 ml**

4 In a stadium, there are sixty rows of ninety seats each. How many seats are there, altogether? **5400**

5 From four, subtract seven. **−3**

6 What is nought point nought eight, multiplied by one hundred? **8**

7 Write one quarter, as a decimal. **0.25**

8 What is the area of a square of sides of length four centimetres? **16 cm²**

9 Write nine hundredths, as a percentage. **9%**

10 Write seventy-five per cent, as a decimal. **0.75**

---

11 A climber has ascended thirty per cent of a cliff of height four hundred and seventy feet. How high has the climber ascended? **141 ft**

12 Multiply three hundred and four by five. **1520**

13 What is seven eighths of sixteen minutes? **14 min**

14 Add fifty-three to six hundred and ninety-eight. **751**

15 On the line AB on your answer sheet, put an X at a point which you estimate to be forty millimetres from point A. **(40 mm)**

16 Increase one hundred and sixty by sixty per cent. **256**

17 Look at the equation on your answer sheet: p equals eighteen, divided by the sum of q and three. Find p, when q is three. **3**

18 A sum of seventy-two pounds is divided in the ratio four to five. How much is the larger share? **£40**

19 The perimeter of a rectangle is twenty metres. The length of the rectangle is six metres. What is its area? **24 m²**

20 What is eighty, divided by eight hundred? **0.1**

---

21 What is four hundred and eighty-seven, added to nine hundred and forty-five? **1432**

22 How many eighteens are there, in nine hundred? **50**

23 What is the total of sixty-six elevens? **726**

24 The mean of three numbers is seventeen. What is the sum of the three numbers? **51**

25 Look at the equation on your answer sheet. What is y, if x is six? **18**

1. How many kilometres are there, in six thousand, eight hundred and twenty metres?   6.82 km
2. What is six thousand, multiplied by seventy?   420000
3. What are thirty-four fours?   136
4. To minus three, add minus two.   −5
5. Write in figures, half a million.   500000
6. What is five point two four, multiplied by one thousand?   5240
7. Write forty-one thousandths, as a decimal.   0.041
8. What is the area of a triangle of height seven metres and length of base four metres?   14 m²
9. Write three fifths, as a decimal.   0.6
10. Write nought point two one, as a fraction. $\frac{21}{100}$

11. The charge for a sun lounger on a beach is three pounds.
There are four hundred and seventeen sun loungers at a resort.
What is the greatest amount of money that could be made in a day?   £1251
12. What is five per cent of nine hundred?   45
13. From three hundred and three, subtract sixty-eight.   235
14. What is half of one hundred and fifty metres?   75 m
15. Look at your answer sheet.
Estimate the size of the angle, in degrees.   (30°)
16. Three angles of a quadrilateral are one hundred degrees, sixty degrees, and one hundred and forty degrees.
What is the size of the fourth angle?   60°
17. There are fourteen beans in a tin.
The size of the tin is increased by fifty per cent.
How many beans will there now be, in a tin?   21
18. What is three hundred and forty, multiplied by forty?   13600
19. Look at the equation on your answer sheet: d equals three times the answer to e minus four.
Find d, when e is thirteen.   27
20. One eighth of a number is forty.
What is the number?   320

21. A box contains twelve tennis balls.
How many boxes can be filled from a sack containing two hundred and forty tennis balls?   20
22. Multiply thirty-four by eleven.   374
23. What is five hundred and eighty-eight, divided by six?   98
24. Find the mean of one, four, five, six, and four.   4
25. Look at the equation on your answer sheet. What is x?   6

1. A temperature of minus three rises to six degrees.
By how many degrees has the temperature risen?   9°
2. How many centimetres are there, in eighty-four millimetres?   8.4 cm
3. Multiply three hundred by six hundred.   180000
4. How many kilograms are there, in five thousand grams?   5 kg
5. Multiply fifty-five by five.   275
6. What is four point one one, divided by one hundred?   0.0411
7. Write seven tenths, as a percentage.   70%
8. What is minus eight, multiplied by minus two?   16
9. What is seventy-three hundredths, written as a decimal?   0.73
10. Write nought point eight, as a percentage.   80%

11. On your answer sheet, write down all the prime numbers between twenty-five and thirty.   29
12. Write one and a third hours, in minutes.   80 min
13. Ten per cent of a number is thirty.
What is the number?   300
14. What is two fifths of fifty-five pounds?   £22
15. Multiply five hundred and four by nine.   4536
16. Look at the equation on your answer sheet: y equals x squared, minus three.
What is y, when x is nine?   78
17. Multiply four hundred by one hundred and seventy.   68000
18. Increase thirty by five sixths.   55
19. Two angles of a triangle are sixty degrees and one hundred degrees.
What is the size of the third angle of the triangle?   20°
20. Write, as a percentage, seven divided by twenty-five.   28%

21. Which number is halfway between five point five and thirteen point five?   9.5
22. Three hundred and forty sweets are divided equally between seventeen children.
How many sweets does each child receive?   20
23. Find the mean of five, three, eight, five, four, and five.   5
24. Add six hundred and seventy-five to four hundred and eighty-seven.   1162
25. What is two squared less than five squared?   21

D27

1 What is the area of a rectangle of length nine metres, and width seven metres. — 63 m²
2 Multiply four hundred by fifty. — 20000
3 From minus three, subtract minus two. — −1
4 How many centimetres are there, in six and a half metres? — 650 cm
5 Approximately how many gallons are there, in eighteen litres? — 4 gal
6 What is one point five seven, multiplied by one thousand? — 1570
7 Write three fifths, as a decimal. — 0.6
8 How many litres are there, in thirty-five thousand millilitres? — 35 litres
9 Write seventy-one thousandths, as a decimal. — 0.071
10 Write forty-seven per cent, as a decimal. — 0.47

11 Look at your answer sheet. Estimate the length of the line AB, in centimetres. — (4 cm)
12 One quarter of a number is four point five. What is the number? — 18
13 Increase six hundred and ninety-four by eighty-five. — 779
14 A man sleeps for one third of the twenty-four hours in a day. For how many hours does he sleep each day? — 8 hrs
15 What is four hundred and eleven, multiplied by six? — 2466
16 The area of a rectangle is twenty-one centimetres squared. The length of the rectangle is seven centimetres. What is its perimeter? — 20 cm
17 Look at the equation on your answer sheet: p equals four q squared. Find p, when q is six. — 144
18 The speed of two boats is in the ratio nine to four. The speed of the second boat is twenty kilometres per hour. What is the speed of the first boat? — 45 km/hr
19 Increase eighteen by one third. — 24
20 What is twenty, divided by four hundred? — 0.05

21 Fifteen workers at a factory are given a bonus of twenty-three pounds each. What is the total of the bonus payments? — £345
22 How many twelves are there, in two hundred and forty? — 20
23 Look at the map on your answer sheet, which has a scale of one centimetre to four kilometres. Estimate the direct distance from point A to point B, in kilometres. — (12 km)
24 The mean of nine numbers is eleven. What is the sum of the eleven numbers? — 99
25 What is nine squared, added to three squared? — 90

D28

1 A temperature of five degrees falls by nine degrees. What is the new temperature? — −4°
2 How many millimetres are there, in ninety centimetres? — 900 mm
3 Multiply ninety by twenty. — 1800
4 How many kilometres are there, in five thousand, seven hundred and thirty metres? — 5.73 km
5 Multiply twenty-four by six. — 144
6 What is twenty, divided by minus four? — −5
7 Write one tenth, as a decimal. — 0.1
8 What is two point nine nine, divided by one hundred? — 0.0299
9 Write nine hundredths, as a percentage. — 9%
10 Write nought point three nine, as a percentage. — 39%

11 Write down all the prime numbers between thirty and thirty-five. — 31
12 What is seventy-seven less than three hundred and two? — 225
13 Look at your answer sheet. Estimate the size of the angle, in degrees. — (25°)
14 What is seven tenths of thirty minutes? — 21 min
15 There are four players in a game of whist. How many players are there, if two hundred and seventeen matches are being played in a tournament? — 868
16 What is four hundred and twenty, multiplied by six hundred? — 252000
17 Reduce twenty-four by one third. — 16
18 Look at the equation on your answer sheet: s equals thirty, minus five t squared. Find s, when t is two. — 10
19 Three angles of a quadrilateral are one hundred and thirty degrees, one hundred and twenty degrees, and eighty degrees. What is the size of the fourth angle? — 30°
20 Increase eighty-one kilograms by one hundred per cent. — 162 kg

21 Which number is halfway between seven point five and thirteen point five? — 10.5
22 How many forties are there, in six hundred and eighty? — 17
23 Find the mean of twelve, eight, four, three, seven, and two. — 6
24 From seven squared, subtract four squared. — 33
25 What is three hundred and fifty-seven, added to nine hundred and fifty-five? — 1312

D29

1 What is the area of a triangle of height seven metres and length of base six metres? 21 m²
2 Multiply fifty-three by seven. 371
3 How many grams are there, in fifty kilograms? 50000 g
4 Multiply eight hundred by two hundred. 160000
5 From eight, subtract nine. −1
6 What is nought point seven eight nine, multiplied by one hundred? 78.9
7 Write fifty-seven hundredths, as a percentage. 57%
8 How many metres are there, in three hundred centimetres? 3 m
9 Write three quarters, as a decimal. 0.75
10 Write twenty-five per cent, as a fraction in its lowest terms. $\frac{1}{4}$

11 Seventy per cent of the one hundred and forty people in a village have been immunised with the flu vaccine. How many people is this? 98
12 From eight hundred and eight, subtract sixty-nine. 739
13 What is four hundred and eighteen, multiplied by three? 1254
14 What is four fifths of forty-five grams? 36 g
15 Write three and three quarters of an hour, in minutes. 225 min
16 What is two hundred and ten, multiplied by five hundred? 105000
17 Seventy per cent of a two hundred and forty pound Christmas budget has been spent. How much money is left? £72
18 The perimeter of a rectangle is thirty-two centimetres. The width of the rectangle is six centimetres. What is its area? 60 cm²
19 Look at the equation on your answer sheet: d equals the square of: e minus three. What is d, when e is twelve? 81
20 Write, as a decimal, nine divided by twenty-five. 0.36

21 The mean of three numbers is twenty-four. What is the sum of the numbers? 72
22 What is two hundred and twenty-four, divided by fourteen? 16
23 What is forty-eight, multiplied by eleven? 528
24 How many eights are there, in seven hundred and eighty-four? 98
25 Look at the equation on your answer sheet. Find y, if x is six. 34

D30

1 How many litres are there, in one thousand, eight hundred millilitres? 1.8 litres
2 Multiply seventy by seventy. 4900
3 How many metres are there, in eighty-three kilometres? 83000 m
4 A temperature of three degrees falls to minus three degrees. By how many degrees has the temperature fallen? 6°
5 Approximately how many miles are equivalent to eighty kilometres? 50 miles
6 Write seventeen hundredths, as a decimal. 0.17
7 What is minus five, multiplied by minus two? 10
8 Write one fifth, as a percentage. 20%
9 What is nought point seven one two, multiplied by one hundred? 71.2
10 Write nought point four, as a percentage. 40%

11 The three angles of a quadrilateral are one hundred and thirty degrees, one hundred and twenty degrees, and eighty degrees. What is the size of the fourth angle? 30°
12 What is forty-eight less than seven hundred and eight? 660
13 What is three quarters of one hundred and sixty? 120
14 What is three hundred and twenty-four, multiplied by five? 1620
15 On the line AB on your answer sheet, put an X at a point which you estimate to be thirty millimetres from point A. (30 mm)
16 A piece of wood of length forty centimetres is cut into two pieces whose lengths are in the ratio of five to three. What is the length of the longest piece? 25 cm
17 Multiply two hundred by six hundred and fifty. 130000
18 Look at the equation on your answer sheet: y is x squared, minus two. What is y, when x is seven? 47
19 Increase eighty by seventy five per cent. 140
20 One third of a number is one point five. What is the number? 4.5

21 Look at the map on your answer sheet, which has a scale of one centimetre to two kilometres. Estimate the direct distance from point A to point B, in kilometres. (5.6 km)
22 Multiply thirty-five by twelve. 420
23 How many sixteens are there, in nine hundred and sixty? 60
24 Find the mean of three, five, four, nine, and four. 5
25 From eight squared, subtract five squared. 39

D31

1. What is the area of a rectangle of length nine metres and width four metres?   36 m²
2. Write in figures: one quarter of a million.   250000
3. What is four hundred, multiplied by eighty?   32000
4. How many millilitres are there, in twenty-six litres?   26000 ml
5. A temperature rises from minus seven degrees to ten degrees. By how many degrees has the temperature risen?   17°
6. How many centimetres are there, in fifteen millimetres?   1.5 cm
7. Write twenty-nine hundredths, as a decimal.   0.29
8. Write seven tenths, as a percentage.   70%
9. What is three point nine seven, divided by one thousand?   0.00397
10. Write nought point one six, as a fraction in its lowest terms. $\frac{4}{25}$

11. There are eight muffins in a bag. How many muffins are there, in five hundred and twenty bags?   4160
12. What is sixty per cent of seven hundred and fifty?   450
13. What is one fifth of seventy-five?   15
14. To one hundred and ninety-three, add sixty-eight.   261
15. Look at your answer sheet. Estimate the length of the line AB, in millimetres.   50 mm
16. The value of an antique watch is increased by two hundred per cent, from fifty-seven pounds. What is the new value of the antique?   £171
17. Look at the equation on your answer sheet: y is thirteen, minus two x. Find y, when x is three.   7
18. What is eight, divided by one thousand, six hundred?   0.005
19. The perimeter of a square is twenty-eight millimetres. What is its area?   49 mm²
20. Reduce forty by three eighths.   25

21. An aeroplane flight costs nine hundred and seventy-six pounds outward, but a standby ticket costs only five hundred and fifty-five pounds on return. What is the total cost of the tickets?   £1531
22. How many fifteens are there, in four hundred and fifty?   30
23. One dozen is twelve. How many is sixteen dozen?   192
24. What is seven squared plus three squared?   58
25. Look at the equation on your answer sheet. Find x.   4

D32

1. How many kilograms are there, in eight thousand, two hundred and sixty grams?   8.26 kg
2. Add minus three to minus three.   −6
3. There are eight men retiring from a company, each one aged sixty-five years. What is their combined age?   520 yrs
4. How many centimetres are there, in eleven metres?   1100 cm
5. Multiply seven thousand by eighty.   560000
6. Write one fifth, as a decimal.   0.2
7. What is ninety-seven point eight, divided by one hundred?   0.978
8. Write sixty-seven hundredths, as a percentage.   67%
9. What is the area of a square of side of length eight centimetres?   64 cm²
10. Write seven per cent, as a decimal.   0.07

11. A box contains five hundred and eighty-seven ball bearings. Another forty-eight ball bearings are added. How many are there, altogether?   635
12. Write one and a third hours, in minutes.   80 min
13. What is two hundred and three, multiplied by seven?   1421
14. What is three eighths of fifty-six?   21
15. What is eighty-five per cent of one hundred?   85
16. Look at the equation on your answer sheet: p equals the square of the sum of q and two. What is p, when q is five?   49
17. What is the total value of one hundred and sixty notes, each of which is a fifty-pound note?   £8000
18. A suitcase has its weight reduced by ten per cent from forty kilograms. What is the new weight of the case?   36 kg
19. The area of a rectangle is fifteen metres squared. The length of the rectangle is five metres. What is its perimeter?   16 m
20. Increase forty by one eighth.   45

21. How many weeks are there, in eight hundred and sixty-eight days?   124 wks
22. Find the mean of three point five and seven point five.   5.5
23. How many nineteens are there, in five hundred and seventy?   30
24. The mean of three numbers is twenty-four. What is the sum of the numbers?   72
25. Look at the equation on your answer sheet. What is y, when x is seven?   36

1   Nine people working at a factory each receive a bonus of forty-one pounds. What was the total bonus earned by these nine workers?   £369
2   From one, subtract four.   −3
3   What is five hundred, multiplied by six hundred?   300000
4   How many millimetres are there, in four point one centimetres?   41 mm
5   How many kilometres are there, in twenty-two thousand metres?   22 km
6   Write nine hundredths, as a decimal.   0.09
7   Write seventy-five per cent, as a fraction. $\frac{3}{4}$
8   What is two point two seven, divided by one hundred?   0.0227
9   Write two fifths, as a percentage.   40%
10   What is minus two, multiplied by minus two?   4

11   A boat has caught one hundred and seventy-seven kilograms of fish. The next net landed has a further eighty-nine kilograms of fish. How much fish is there, altogether?   266 kg
12   What is twenty-five per cent of four hundred kilograms?   100 kg
13   Multiply five hundred and eleven by nine.   4599
14   On your answer sheet, draw an arrow on the protractor scale to indicate your estimate of a sixty-degree angle.   (60°)
15   What is nine tenths of seventy?   63
16   An alloy contains iron and copper in the ratio of five to one. A block of this alloy contains forty kilograms of iron. What weight of copper does it contain?   8 kg
17   One fifth of a number is twenty. What is the number?   100
18   Look at the equation on your answer sheet: c is double the answer to d add three. Find c, when d is ten.   26
19   Write, as a percentage, twenty-one divided by twenty-five.   84%
20   The two angles of a triangle are eighty-five degrees, and eighty degrees. What is the size of the third angle?   15°

21   Which number is halfway between six point five and sixteen point five?   11.5
22   How many fourteens are there, in four hundred and twenty?   30
23   Find the mean of five, six, two, four, and three.   4
24   Add eight hundred and eighty-seven to six hundred and fifty-nine.   1546
25   What is four squared and two squared?   20

1   How many kilometres are there, in five thousand, four hundred metres?   5.4 km
2   What is six hundred, multiplied by sixty?   36000
3   A temperature of minus two falls by two degrees. What is the new temperature?   −4°
4   How many millilitres are there, in four point nine litres?   4900 ml
5   Approximately how many kilometres are equivalent to twenty miles?   32 km
6   What is the area of a rectangle of length nine metres and width eight metres?   72 m²
7   Write sixty-one hundredths, as a percentage.   61%
8   What is three point nought nine one, multiplied by one hundred?   309.1
9   Write four fifths, as a decimal.   0.8
10   Write nought point four six, as a percentage.   46%

11   One fifth of a 35-litre tank of petrol has been used on a journey. How many litres of petrol have been used?   7 litres
12   From three hundred and nine, subtract seventy-four.   235
13   Write down all the prime numbers between thirty-five and forty.   37
14   What is four hundred and four, multiplied by eight?   3232
15   Look at the answer sheet. Estimate the length of the line AB in millimetres.   (30 mm)
16   Write, as a decimal, seven divided by twenty-five.   0.28
17   Look at the equation on your answer sheet: a is two c squared. Find a, when c is six.   72
18   What is thirty, divided by three thousand?   0.01
19   The contents of a box of seventy matches is increased by ninety per cent. How many matches are in the new box?   133
20   The three angles of a quadrilateral are ninety degrees, one hundred and thirty degrees, and seventy degrees. What is the size of the fourth angle?   70°

21   A man has received tax bills for three hundred and eighty-six pounds, and seven hundred and thirty-five pounds. What is the total amount to be paid?   £1121
22   What is twelve, multiplied by fifteen?   180
23   From seven squared, subtract six squared.   13
24   How many twelves are there, in eight hundred and forty?   70
25   Look at the map on your answer sheet, which has a scale of one centimetre to ten kilometres. Estimate the direct distance from point A to point B, in kilometres.   (22.5 km)

D35

Answers

1 A temperature of four degrees falls by seven degrees.
What is the new temperature? −3°
2 Multiply five thousand by eighty. 400000
3 How many kilometres are there, in seventy metres? 0.07 km
4 Multiply sixty-one by three. 183
5 How many centimetres are there, in twenty-six millimetres? 2.6 cm
6 Write one quarter, as a percentage. 25%
7 What is nought point nine one five, multiplied by one hundred? 91.5
8 What is twelve, divided by minus three? −4
9 Write thirty hundredths, as a percentage. 30%
10 Write eighty-three per cent, as a decimal. 0.83

11 A music centre normally costing eight hundred and eleven pounds has been reduced in price by fifty-seven pounds.
What is the new price? £754
12 What is five eighths of four thousand? 2500
13 Twenty-five per cent of a number is one hundred.
What is the number? 400
14 Multiply three hundred and thirty-two by six. 1992
15 On the line AB on your answer sheet, put an X at a point which you estimate to be two centimetres from point A. (2 cm)
16 A sixty-kilogram weight is reduced by five per cent.
To what is the weight reduced? 57 kg
17 Look at the equation on your answer sheet: y is x squared, minus four.
Find y, when x is twelve. 140
18 What is five hundred and forty, multiplied by five hundred? 270000
19 The perimeter of a rectangle is thirty-four millimetres.
The length of the rectangle is nine millimetres.
What is its area? 72 mm²
20 Increase sixty-four by one eighth. 72

21 The mean of five numbers is nine.
What is the sum of the five numbers? 45
22 How many elevens are there, in three hundred and thirty? 30
23 Divide seven hundred and seventy-six by eight. 97
24 What are forty-four elevens? 484
25 Look at the equation on your answer sheet. What is x? 8

D36

Answers

1 How many kilograms are there, in nine thousand, seven hundred and eighty grams? 9.78 kg
2 A temperature of minus four degrees rises to five degrees. By how many degrees has the temperature risen? 9°
2 How many centimetres are there, in five and a half metres? 550 cm
4 What is nine thousand, multiplied by eighty? 720000
5 Write in figures: three quarters of a million. 750000
6 Write two fifths, as a decimal. 0.4
7 What is the area of a triangle of height seven metres and base of length eight metres? 28 m²
8 Write eighty-nine hundredths, as a percentage. 89%
9 What is eight point eight one, divided by one hundred? 0.0881
10 Write nought point one six, as a fraction in its lowest terms. $\frac{4}{25}$

11 Sarah has driven three tenths of her seventy-kilometre journey.
How far has she actually driven? 21 km
12 What is eighty per cent of two hundred and ten? 168
13 Multiply three hundred and twelve by seven. 2184
14 Write two and three quarters of an hour, in minutes. 165 min
15 Add forty-eight to four hundred and eighty-three. 531
16 Multiply two hundred and sixty by five hundred. 130000
17 Three angles of a quadrilateral are seventy degrees, one hundred degrees, and one hundred and ten degrees.
What is the size of the fourth angle? 80°
18 A tank is partly full with fifty litres of petrol. More petrol is added, increasing the amount of petrol by nine tenths.
How many litres of petrol are now in the tank? 95 litres
19 An amount of one hundred and ten pounds is divided between Ali and Ben in the ratio of six to five.
How much does Ben receive? £50
20 Look at the equation on your answer sheet: a is equal to twenty-four, divided by the sum of c and three.
Find a, when c is nine. 2

21 Add these numbers: six hundred and forty-five and eight hundred and seventy-five. 1520
22 How many twelves are there, in seven hundred and twenty? 60
23 Find the mean of five, one, nine, four, and six. 5
24 Multiply forty-four by eleven. 484
25 What is nine squared plus three squared? 90

79

D37

1 A drum can hold ten gallons. Approximately how many litres are there, in ten gallons? — 45 litres

2 How many litres are there, in nine thousand millilitres? — 9 litres

3 Subtract six from three. — −3

4 What is ninety-one, multiplied by four? — 364

5 Multiply seven hundred by fifty. — 35000

6 Write forty-nine hundredths, as a percentage. — 49%

7 What is four point eight three, multiplied by one thousand? — 4830

8 Write thirty-five per cent, as a fraction in its lowest terms. — $\frac{7}{20}$

9 What is minus sixteen, divided by minus four? — 4

10 Write three tenths, as a decimal. — 0.3

11 Fifty per cent of a herd of eighty-four sheep have been sheared. How many sheep have been sheared? — 42

12 What is one quarter of three hundred and sixty grams? — 90 g

13 From two hundred and twenty-one, subtract sixty-four. — 157

14 What is four hundred and twelve, multiplied by five? — 2060

15 On the line AB on your answer sheet, put an X at a point which you estimate to be forty millimetres from point A. — (40 mm)

16 Write, as a percentage, eleven divided by twenty-five. — 44%

17 The perimeter of a square is twenty-four metres. What is the area of the square? — 36 m²

18 Look at the equation on your answer sheet: d equals four times the answer to e subtract two. Find d, when e is eight. — 24

19 One sixth of a number is two point five. What is the number? — 15

20 Reduce a waiting list of fifty-six people, by three quarters. — 14

21 Which number is exactly halfway between five point five and eleven point five? — 8.5

22 How many elevens are there, in four hundred and forty? — 40

23 Find the mean of six, five, two, and seven. — 5

24 What is eight hundred and seventy-six, divided by six? — 146

25 To two cubed, add five squared. — 33

D38

1 What is the area of a rectangle of length eight metres and width four metres? — 32 m²

2 Multiply six thousand by eight hundred. — 4800000

3 What are fifteen fives? — 75

4 A temperature of five degrees falls by nine degrees. What is the new temperature? — −4°

5 How many metres are there, in seven kilometres? — 7000 m

6 What is nought point nought nine five, multiplied by one hundred? — 9.5

7 Write seven hundredths, as a decimal. — 0.07

8 How many centimetres are there, in sixty millimetres? — 6 cm

9 Write four fifths, as a percentage. — 80%

10 Write nought point one two, as a fraction in its lowest terms. — $\frac{3}{25}$

11 Nigel has eaten four fifths of a 400-calorie meal. How many calories has he eaten so far? — 320 cal

12 Ten per cent of a number is seventy. What is the number? — 700

13 To three hundred and ninety-three, add forty-eight. — 441

14 What is four hundred and fourteen, multiplied by three? — 1242

15 On your answer sheet, draw an arrow on the protractor scale to indicate your estimate of a forty-five degree angle. — (45°)

16 Two lengths of wood are in the ratio five to seven. The longer length is one hundred and twelve millimetres. What is the shorter length? — 80 mm

17 Look at the equation on your answer sheet: y is equal to six x squared. Find y, when x is eight. — 384

18 Reduce eighteen by two thirds. — 6

19 Two angles of a triangle are both eighty degrees. What is the size of the third angle of the triangle? — 20°

20 What is forty, divided by four hundred? — 0.1

21 Three hundred and ninety chairs are to be arranged in rows of thirteen chairs each. How many rows will there be? — 30

22 Find the mean of five point five and seven point five. — 6.5

23 Take four squared from five squared. — 9

24 The annual cost of a company's two mobile phones are eight hundred and ninety-six pounds, and four hundred and forty-six pounds. What is the total cost? — £1342

25 The mean of four numbers is twenty-one. What is the sum of the four numbers? — 84

D39

Answers

1. How many metres are there, in three hundred and ten centimetres? — 3.1 m
2. To minus three, add minus six. — –9
3. What is fifty-nine, multiplied by six? — 354
4. How many grams are there, in eighteen kilograms? — 18000 g
5. What is six thousand, multiplied by seventy? — 420000
6. Write nine tenths, as a decimal. — 0.9
7. What is the area of a square with sides of length six centimetres? — 36 cm²
8. What is two point two seven, multiplied by one hundred? — 227
9. Write sixty-six hundredths, as a percentage. — 66%
10. Write three per cent, as a decimal. — 0.03

11. Two items of furniture cost two hundred and eighty-seven pounds, and forty-seven pounds. What is the total cost? — £334
12. What is forty per cent of nine hundred and fifty? — 380
13. What is two hundred and seven, multiplied by eight? — 1656
14. What is one third of fifty-four grams? — 18 g
15. On the line AB on your answer sheet, put an X at a point which you estimate to be two centimetres from point A. — (2 cm)
16. Multiply eight thousand, one hundred by fifty. — 405000
17. A dress normally costing sixty-four pounds has its price reduced by twenty-five per cent in a sale. What is the sale price? — £48
18. Look at the equation on your answer sheet: d is the square of the sum of e and three. What is d, when e is seven? — 100
19. Reduce eighteen by one ninth. — 16
20. The area of a rectangle is twenty-four centimetres squared. The width of the rectangle is four centimetres. What is its perimeter? — 20 cm

21. Look at the map on your answer sheet, which has a scale of one centimetre to five kilometres. Estimate the direct distance between point A and point B, in kilometres. — (16 km)
22. Multiply thirty-one by twelve. — 372
23. How many twelves are there, in four hundred and eighty? — 40
24. What is the difference between nine squared and eight squared? — 17
25. Look at the equation on your answer sheet. What is y, when x is four? — 33

D40

Answers

1. A temperature of two degrees falls by three degrees. What is the new temperature? — –1°
2. Multiply eighty by seventy. — 5600
3. What is seventy-two, multiplied by seven? — 504
4. How many grams are there, in two point two kilograms? — 2200 g
5. Approximately how many pints are there, in ten litres? — 17.5 pints
6. Write one tenth, as a percentage. — 10%
7. What is nought point six seven two, multiplied by one hundred? — 67.2
8. Write forty-seven hundredths, as a decimal. — 0.47
9. Write forty-five per cent, as a fraction in its lowest terms. — $\frac{9}{20}$
10. Minus three, multiplied by minus two. — 6

11. A man has spent seven eighths of the four thousand dollars he brought with him into the country. How much has he spent, in dollars? — $3500
12. Write two and two thirds hours, in minutes. — 160 min
13. From nine hundred and seven, subtract thirty-eight. — 869
14. What is ten per cent of eight hundred and thirty? — 83
15. What is five hundred and eleven, multiplied by nine? — 4599
16. The three angles of a quadrilateral are ninety degrees, one hundred and ten degrees, and ninety degrees. What is the size of the fourth angle? — 70°
17. Multiply three thousand, two hundred by two hundred. — 640000
18. Look at the equation on your answer sheet: y is equal to x squared, plus three. Find y, when x is six. — 39
19. The cost of a bag in New York is increased by one ninth from forty-five dollars. What is the new cost of the bag? — $50
20. Four pounds fifty is divided between Colin and Derek in the ratio five to four. How much does Colin receive? — £2.50

21. A prize of eight hundred and sixty-four pounds is to be divided equally between nine people. How much does each person receive? — £96
22. Find the mean of one, five, eight, ten, and eleven. — 7
23. Multiply fifteen by eleven. — 165
24. How many elevens are there, in five hundred and fifty? — 50
25. Look at the equation on your answer sheet. What is y, when x is eight? — 58

1 The length of a spanner has been measured as thirteen point eight centimetres, to the nearest millimetre. What is the minimum length it could be?   13.75 cm

2 Write ninety-one per cent, as a fraction. $\frac{91}{100}$

3 Write three fifths, as a decimal.   0.6

4 What is two point eight seven, multiplied by one hundred?   287

5 Multiply minus five by minus five.   25

6 Write three hundredths, as a percentage.   3%

7 What is one half of one third? $\frac{1}{6}$

8 Divide nought point nine by nought point nought one.   90

9 Write forty hundredths, as a decimal.   0.04

10 Multiply nought point nought one by nought point nought nine.   0.0009

11 The perimeter of a rectangle is sixteen centimetres. The length of the rectangle is five centimetres. What is its area?   15 cm²

12 Write, as a percentage, nine, divided by fifty.   18%

13 Multiply two thousand, two hundred by forty.   88000

14 Look at the equation on your answer sheet: C equals three x, taken from twenty-five. Find C, when x is five.   10

15 A forty-pound bill is to be reduced by three tenths. What will the bill be reduced to?   £28

16 Write down, as a whole number, an estimate for the value of nine hundred and eighty-five, divided by thirty-seven.   (25)

17 Multiply forty by nought point four.   16

18 A car is travelling at an average speed of sixty kilometres per hour. How long does it take for the car to travel one hundred and thirty-five kilometres?   2 hrs 15 min

19 Look at the expression on your answer sheet. What is the greatest whole number x can be?   9

20 How many five thousands are there, in one million?   200

21 Look at the equation on your answer sheet: y equals five x, plus nine. What is y, when x is six?   39

22 What is four squared, added to two squared?   20

23 The mean of six numbers is fourteen. What is the sum of the six numbers?   84

24 How many nought point nought fives are there, in four hundred?   8000

25 Look at the expression on your answer sheet. Work out an approximate answer to this calculation.   15

1 A rectangle is of length nine centimetres and width seven centimetres. What is its area?   63 cm²

2 Write three tenths, as a decimal.   0.3

3 What is eight point one, divided by one thousand?   0.0081

4 Write seventy-nine hundredths, as a percentage.   79%

5 Write four per cent, as a decimal.   0.04

6 What is one half of two sevenths? Give your answer as a fraction, in its lowest terms. $\frac{1}{7}$

7 The height of a clock has been measured as twelve point four centimetres, measured to the nearest millimetre. What is the minimum height the clock could be?   12.35 cm

8 Multiply nought point nought four by nought point nought nought two.   0.00008

9 What is one third of three fifths? Give your answer as a fraction, in its lowest terms. $\frac{1}{5}$

10 Write six hundred thousandths, as a decimal.   0.6

11 Two angles of a triangle are each sixty-five degrees. What is the size of the third angle?   50°

12 The ratio of the number of cats to the number of dogs owned by the children in a year group is five to three. There are sixty cats. How many dogs are there?   36

13 Reduce sixty-four by seventy-five per cent.   16

14 Look at the equation on your answer sheet: d is sixteen, divided by the sum of e and three. What is d, when e is five?   2

15 Divide thirty by six hundred. Give your answer as a decimal.   0.05

16 Write down, as a decimal to one decimal place, an estimate for the value of forty-one, divided by forty-nine.   0.8

17 Look at the expression on your answer sheet. What is the smallest whole number x can be?   4

18 Multiply nought point nought eight by sixty.   4.8

19 Thirty-two multiplied by twenty-one is six hundred and seventy-two. What is six hundred and seventy-two, divided by three point two?   210

20 How long does it take to drive two hundred and sixty kilometres, at an average speed of eighty kilometres per hour?   3 hrs 15 min

21 Look at the equation on your answer sheet: y equals seven x, plus nine. What is y, when x is six?   51

22 What is seven squared, plus five squared?   74

23 Look at the equation on your answer sheet: eight x, minus fifteen, equals seventeen. What is x?   4

24 How many nought point nought fours are there in eight hundred?   20000

25 Look at the expression on your answer sheet. Work out an approximate answer to this calculation.   18

E3 — Answers

1. Find the area of a triangle which has a base of length eight centimetres and a height of six centimetres. — 24 cm²
2. Write twenty-three hundredths, as a percentage. — 23%
3. Write nought point eight five as a fraction, in its lowest terms. — $\frac{17}{20}$
4. Write one quarter, as a decimal. — 0.25
5. Divide fifteen point two by one thousand. — 0.0152
6. The width of a finger has been measured as twenty-four millimetres, to the nearest millimetre. What is the maximum width the finger could be? — 24.5 mm
7. Multiply nought point nought nought two by nought point nought two. — 0.00004
8. Write eighty thousandths, as a decimal. — 0.08
9. What is one quarter of one sixth? — $\frac{1}{24}$
10. Divide nought point nought eight by nought point one. — 0.8

11. Look at the equation on your answer sheet: p equals the square of the sum of q and three. Find p, when q is five. — 64
12. Write, as a decimal, twenty-one, divided by twenty-five. — 0.84
13. Ten per cent of a batch of thirty silicon chips are faulty. How many have been correctly manufactured? — 27
14. Three angles of a quadrilateral are seventy degrees, eighty degrees, and ninety degrees. What is the size of the fourth angle? — 120°
15. Multiply seven thousand, five hundred by forty. — 300000
16. Multiply eight hundred and sixty by seventy. — 60200
17. Estimate, to the nearest whole number, the value of nine per cent of three hundred and fifty-one. — 32 or 35
18. Multiply nine hundred by nought point five. — 450
19. A woman made a journey of thirty kilometres in twenty minutes. What was her average speed? — 90 km/hr
20. Look at the expression on your answer sheet. What is the smallest whole number x can be? — 7

21. Look at the equation on your answer sheet: y equals six x, minus eleven. What is y, when x is seven? — 31
22. The mean of three numbers is twenty-three. What is the sum of the three numbers? — 69
23. Add three squared to two cubed. — 17
24. Look at the expression on your answer sheet. Work out an approximate answer to this calculation. — (400)
25. How many nought point nought twos are there in ten? — 500

E4 — Answers

1. What is nought point nought nine seven multiplied by one hundred? — 9.7
2. Divide eight by minus two. — −4
3. Write three fifths, as a percentage. — 60%
4. Write fifty-three hundredths, as a decimal. — 0.53
5. Write nought point four nine, as a percentage. — 49%
6. What is one half of nine tenths? — $\frac{9}{20}$
7. Write two hundred ten-thousandths, as a decimal. — 0.02
8. The length of a path has been measured as eight point four two metres, to the nearest centimetre. What is the minimum length the path could be? — 8.415 m
9. What is one third of one quarter? — $\frac{1}{12}$
10. Divide nought point nought three by nought point nought nought one. — 30

11. The area of a rectangle is forty metres squared. The width of the rectangle is four metres. What is the perimeter? — 28 m
12. Multiply four hundred and ninety by thirty. — 14700
13. Increase twenty by five per cent. — 21
14. Look at the equation on your answer sheet: y equals three times the answer to x, take away two. Find y, when x is nine. — 21
15. An amount of sixty-three pounds is to be increased by eight ninths. What will the amount then be? — £119
16. Multiply eight hundred and seventy by ninety. — 78300
17. Divide three million by thirty thousand. — 100
18. Write down, as a whole number, an estimate for the value of three hundred and eighty-one, divided by two hundred and twenty-one. — (2)
19. Look at the expression on your answer sheet. What is the smallest whole number x can be? — 8
20. An object travels at a constant speed of thirty metres per second for six and a half seconds. How far has the object moved in this time? — 195 m

21. Look at the equation on your answer sheet: seven x, minus fifteen, equals twenty-seven. What is x? — 6
22. What is seven squared, minus two squared? — 45
23. Look at the equation on your answer sheet: y equals six x, minus fourteen. What is y, when x is nine? — 40
24. How many nought point nought fours are there, in one hundred and sixty? — 4000
25. Look at the expression on your answer sheet. Work out an approximate answer to this calculation. — (90)

E5
Answers

1. Find the area of a rectangle of length seven centimetres and width six centimetres. — 42 cm²
2. Write one fifth, as a decimal. — 0.2
3. Write nineteen hundredths, as a percentage. — 19%
4. Write thirty-five per cent as a fraction, in its lowest terms. — $\frac{7}{20}$
5. Multiply nought point four nought seven by one thousand. — 407
6. The width of a classroom has been measured as fourteen point seven metres, to the nearest tenth of a metre. What is the minimum width the classroom could be? — 14.65 m
7. What is nought point two, multiplied by nought point nought two, multiplied by nought point two? — 0.0008
8. Write ten thousandths, as a decimal. — 0.01
9. What is nought point nought nought three, divided by nought point nought one? — 0.3
10. What is one half of one tenth? — $\frac{1}{20}$

11. A sixty-centimetre length of ribbon is to be reduced by five twelfths. To what length is the ribbon to be reduced? — 35 cm
12. What is four thousand, three hundred, multiplied by five hundred? — 2150000
13. Look at the equation on your answer sheet: d equals three e squared. Find d, when e is eight. — 192
14. Two angles of a triangle are ninety degrees and seventy degrees. What is the size of the third angle? — 20°
15. Thirty millimetres of water is divided between two containers in the ratio of two to one. What is the largest share? — 20 ml
16. A man runs six kilometres in forty minutes. What is the average speed of the runner? Give your answer in kilometres per hour. — 9 km/hr
17. How many five thousands are there, in two million? — 400
18. Look at the expression on your answer sheet. What is the greatest whole number x can be? — 14
19. Multiply nought point nought three by seventy. — 2.1
20. Estimate the value of fifty-one per cent of two hundred and one. — (102 or 100)

21. Write down on your answer sheet the sum of six squared and three squared. — 45
22. Find the mean of four, seven, six, and eight. — 6.25
23. Look at the equation on your answer sheet: twenty-one equals thirty-seven, minus four x. What is x? — 4
24. An object is moving at a speed of forty-eight metres per minute. Write this speed in metres per second. — 0.8 m/s
25. Look at the expression on your answer sheet. Work out an approximate answer to this calculation. — (0.2)

E6
Answers

1. The height of a cupboard has been measured to two point three four metres, to the nearest centimetre. What is the maximum height, the cupboard could be? — 2.345 m
2. What is minus three, multiplied by minus three? — 9
3. Write three tenths, as a percentage. — 30%
4. What is sixty one point one, divided by one hundred? — 0.611
5. Write nought point four eight as a fraction, in its lowest terms. — $\frac{12}{25}$
6. What is nought point six, divided by nought point nought three? — 20
7. What is one quarter of one third? — $\frac{1}{12}$
8. Write thirteen hundredths, as a decimal. — 0.13
9. Write four hundred thousandths, as a decimal. — 0.4
10. Multiply nought point nought nought two by nought point nought three. — 0.00006

11. The three angles of a quadrilateral are eighty degrees, eighty degrees, and seventy degrees. What is the fourth angle? — 130°
12. Increase forty-eight by twenty-five per cent. — 60
13. Divide ninety by nine hundred. — 0.1
14. A jacket costing sixty pounds has been reduced by two thirds in a closing-down sale. What is the sale price of the jacket? — £20
15. Look at the equation on your answer sheet: a equals twelve, divided by the sum of c and two. Find a, when c is four. — 2
16. Write down, as a whole number, an estimate for the value of two hundred and eighty-one, divided by fifty-seven. — (5)
17. Look at the expression on your answer sheet. What is the greatest whole number x can be? — 7
18. Multiply six hundred by ninety-two. — 55200
19. Sixty-eight multiplied by thirty-seven equals two thousand, five hundred and sixteen. What is two thousand, five hundred and sixteen, divided by six point eight? — 370
20. How long does it take an object to travel one hundred kilometres, at a speed of thirty kilometres per hour? — 3 hrs 20 min

21. Estimate the value of seventy-eight per cent of two hundred and ninety-seven. — 240
22. What is nine squared, plus six squared? — 117
23. Look at the equation on your answer sheet: y equals nine x, minus fourteen. What is y, when x is five? — 31
24. How many nought point nought sixes are there, in six hundred? — 10000
25. Look at the expression on your answer sheet. Work out an approximate answer to this calculation. — (16)

1   A letter has been found to be of weight eighty-five grams, to the nearest gram. What is the minimum weight the letter could be?   84.5 g

2   What is nought point seven six one, multiplied by one thousand?   761

3   Write one tenth, as a decimal.   0.1

4   Write seventy-three hundredths, as a percentage.   73%

5   Write sixty-five per cent as a fraction, in its lowest terms. $\frac{13}{20}$

6   What is nought point nought nought one, multiplied by nought point nought nought nought one?   0.0000001

7   Find the area of a square of side of length nine centimetres.   81 cm²

8   Divide nought point nought five by nought point nought one.   5

9   Write thirty ten-thousandths, as a decimal.   0.003

10 What is one fifth, divided by one third? $\frac{3}{5}$

11 The perimeter of a square is sixteen centimetres. What is the area of the square?   16 cm²

12 Multiply six hundred and twenty by five hundred.   310000

13 Write, as a decimal, seven, divided by fifty.   0.14

14 Look at the equation on your answer sheet: y equals x squared, plus four. Find y, when x is seven.   53

15 Decrease fifty-eight by fifty per cent. 29

16 What is nought point two, multiplied by six hundred?   120

17 Look at the expression on your answer sheet. What is the smallest whole number x can be?   5

18 A car travels for three and two thirds of an hour, at an average speed of ninety kilometres per hour. How far will the car have travelled?   330 km

19 Write down, as a decimal to one decimal place, an estimate for the value of five hundred and four, divided by one thousand, five hundred and four.   0.3

20 How many two thousands are there, in four million?   2000

21 Two dice are rolled and their scores added together. What is the probability of a total score of at least ten? $\frac{1}{6}$

22 Find the mean of two, seven, nine, and six.   6

23 Look at the equation on your answer sheet: seven x, minus thirteen, equals thirty-six. What is x?   7

24 From eight squared, subtract seven squared.   15

25 Look at the expression on your answer sheet. Work out an approximate answer to this calculation.   (5)

1   Find the area of a triangle of base of length six centimetres and height seven centimetres.   21 cm²

2   What is nought point one six, multiplied by one hundred?   16

3   Write eleven hundredths, as a decimal.   0.11

4   Write three quarters, as a percentage.   75%

5   Write nought point three four, as a percentage.   34%

6   What is one fifth of three tenths? $\frac{3}{50}$

7   Write eighty hundredths, as a decimal.   0.8

8   What is nought point nought four, divided by nought point nought nought one?   40

9   The capacity of a jug has been measured as nought point four two litres, to the nearest hundredth of a litre. What is the minimum capacity the jug could be?   0.415 litres

10 What is nought point nought four, multiplied by nought point nought nought two?   0.00008

11 Two angles of a triangle are sixty degrees and eighty degrees. What is the size of the third angle?   40°

12 Increase fifty by sixty per cent.   80

13 Look at the equation on your answer sheet: p equals twenty, minus four q. Find p, when q is three.   8

14 What is two thousand, four hundred, multiplied by twenty?   48000

15 The ratio of the perimeter of the triangle to its shortest side is ten to three. The perimeter is forty centimetres. What is the length of the shortest side? 12 cm

16 Estimate the value of thirty-nine per cent of three hundred and ninety-seven pounds.   (£160)

17 Look at the expression on your answer sheet. What is the smallest whole number x can be?   1

18 How many thousands are there, in four million?   4000

19 A salesman drives for three hours and twenty minutes, and covers a distance of three hundred and ninety kilometres. What was the salesman's average speed?   117 km/hr

20 A sheet of card has a thickness of nought point nought eight centimetres. What will be the height, in centimetres, of a pile of two hundred sheets of card? 16 cm

21 Look at the expression on your answer sheet. Work out an approximate answer to this calculation.   (2)

22 Look at the equation on your answer sheet: y equals forty-four, minus six x. What is y, when x is six?   8

23 What is five squared, plus four squared? 41

24 Find the mean of three, five, four, nine, and four.   5

25 How many nought point nought fives are there in five hundred?   10000

1 A parcel has been weighed as one point two kilograms, to the nearest tenth of a kilogram. What is the maximum weight the parcel could be?   1.25 kg
2 What is nine point five nought five, divided by one hundred?   0.09505
3 What is eight, divided by minus four?   −2
4 Write thirty-nine hundredths, as a decimal.   0.39
5 Write three point seven five, as a percentage.   375%
6 What is one sixth, divided by one fifth? $\frac{5}{6}$
7 Write two fifths, as a percentage.   40%
8 Write three hundred thousandths, as a decimal.   0.3
9 What is nought point six, divided by nought point nought one?   60
10 What is one half of one fifth? $\frac{1}{10}$

11 Look at the equation on your answer sheet: d equals double the answer to e plus three.
Find d, when e is thirteen.   32
12 Write, as a percentage, three, divided by twenty-five.   12%
13 Reduce sixty pence by eleven twelfths.   5p
14 The perimeter of a rectangle is twenty-eight centimetres. The width of the rectangle is four centimetres. What is its area?   40 cm²
15 What is fifty, divided by five thousand?   0.01
16 How long does it take to travel one hundred and sixty-five kilometres, at a speed of sixty kilometres per hour?   2 hrs 45 min
17 Multiply thirty by nought point nought six.   1.8
18 Write down, as an integer, an estimate for the value of nine hundred and three, divided by forty-nine.   (18)
19 Look at the expression on your answer sheet. What is the greatest whole number x can be?   5
20 How many thousands are there, in three million?   3000

21 Look at the expression on your answer sheet. Work out an approximate answer to this calculation.   (2)
22 Estimate the value of ninety-one per cent of seven hundred and three.   (630)
23 Look at the equation on your answer sheet: six x, plus nine, equals fifty-seven. What is x?   8
24 How many nought point nought threes are there in one thousand, two hundred?   40000
25 What is two cubed, added to four squared?   24

1 Find the area of a rectangle of length eight centimetres and width seven centimetres.   56 cm²
2 Write forty-nine hundredths, as a percentage.   49%
3 What is nought point three seven two, multiplied by one thousand?   372
4 Write three quarters, as a decimal.   0.75
5 Write nought point three five as a fraction, in its lowest terms. $\frac{7}{20}$
6 What is nought point nought nought three, multiplied by nought point nought two?   0.00006
7 The length of a ribbon has been measured as two point nine five metres, to the nearest centimetre. What is the minimum length the ribbon could be?   2.945 m
8 Write sixty thousandths, as a decimal.   0.06
9 Divide one seventh by one third. $\frac{3}{7}$
10 What is nought point nought five, divided by nought point nought one?   5

11 A workforce of sixteen is to be increased by one hundred per cent.
How many workers will there then be?   32
12 What is six hundred and sixty, multiplied by forty?   26400
13 A length of eighty-eight millimetres is to be reduced by one eighth. What will the reduced length be?   77 mm
14 Three angles of a quadrilateral are one hundred degrees, ninety degrees, and one hundred degrees.
What size is the fourth angle?   70°
15 Look at the equation on your answer sheet. What is y, when x is three?   4
16 Seven hundred workers each receive an eighty-six-pound allocation of shares. What is the cost of the total share allocation?   £60200
17 A car travels at an average speed of eighty kilometres per hour for two hours and forty-five minutes. How far does the car travel in this time?   220 km
18 Write down, as a decimal to one decimal place, an estimate for the value of fifty-seven, divided by seventy-two.   (0.8)
19 Look at the expression on your answer sheet. What is the smallest whole number x can be?   3
20 Seventy-three multiplied by thirty-seven is two thousand, seven hundred and one. What is two thousand, seven hundred and one, divided by nought point three seven?   7300

21 Look at the expression on your answer sheet. Work out an approximate answer to this calculation.   (400)
22 Find the mean of eight, twelve, eight, and sixteen.   11
23 Look at the equation on your answer sheet. What is y, when x is seven?   24
24 What is three squared, taken from six squared?   27
25 An object is moving at a speed of nought point three metres per second. Write this speed in metres per minute.   18 m/min

1 The length of a ribbon has been measured as eighty-four point three centimetres, to the nearest millimetre. What is the minimum length, the ribbon could be?   84.25 cm

2 What is two fifths, as a decimal?   0.4

3 Divide fifteen point four by one thousand.   0.0154

4 Write thirty-three hundredths, as a percentage.   33%

5 Write eighty-three per cent, as a decimal.   0.83

6 What is ten, divided by minus two?   −5

7 Write twenty hundredths, as a decimal.   0.2

8 What is one third of five eighths? $\frac{5}{24}$

9 Divide nought point nine by nought point nought three.   30

10 What is nought point one, multiplied by nought point nought one, multiplied by nought point nought nought one?   0.000001

11 An hour is divided into two lessons, in the ratio of two to one. How many minutes is there in the longer of the two lessons?   40 min

12 What is four thousand, one hundred, multiplied by eighty?   328000

13 Look at the equation on your answer sheet. What is y, when x is four?   80

14 The perimeter of a square is twenty millimetres. What is the area of the square?   25 mm²

15 An amount of twenty pounds is to be increased by four fifths. What will be the increased amount?   £36

16 How many twenty-five thousands are there, in one million?   40

17 A bead weighs nought point nought seven grams. What will be the weight of eighty such beads?   5.6 g

18 Estimate the value of seventy-one per cent of four hundred and five.   (280)

19 A driver covers two hundred kilometres in exactly two and a half hours. What is the driver's average speed for the journey?   80 km/hr

20 Look at the expression on your answer sheet. What is the smallest whole number x can be?   6

21 The mean of six numbers is calculated to be eleven. What is the sum of the six numbers?   66

22 Take four squared from nine squared.   65

23 Look at the equation on your answer sheet: eight x, minus sixteen, equals twenty-four. What is x?   5

24 How many nought point nought threes are there, in three hundred?   10000

25 Look at the expression on your answer sheet. Work out an approximate answer to this calculation.   (20)

1 What is the area of a triangle with base of length six centimetres and height four centimetres?   12 cm²

2 Write one tenth, as a percentage.   10%

3 What is seventeen point nine, divided by one thousand?   0.0179

4 Write forty-three hundredths, as a decimal.   0.43

5 Write nought point six seven, as a percentage.   67%

6 The weight of a football has been measured to be nought point nine six kilograms, to the nearest hundredth of a kilogram. What is the maximum weight the football could be?   0.965 kg

7 What is nought point nought nought three, multiplied by nought point two?   0.0006

8 Write two thousand ten-thousandths, as a decimal.   0.02

9 What is nought point nought seven, divided by nought point nought nought seven?   10

10 What is half of one sixth? $\frac{1}{12}$

11 Two angles of a triangle are sixty-five degrees and eighty-five degrees. What is the size of the third angle of the triangle?   30°

12 Decrease three hundred by one per cent.   297

13 Look at the equation on your answer sheet: a equals c squared, plus five. Find a, when c is five.   30

14 Multiply one hundred and eighty by thirty.   5400

15 Increase forty-five dollars by two ninths. $55

16 Multiply nought point nine by three hundred.   270

17 Write down, as a whole number, an estimate for the value of four hundred and ninety-one, divided by forty-seven.   (10)

18 How long does it take to drive three hundred and fifteen kilometres, at an average speed of seventy kilometres per hour?   4 hrs 30 min

19 Look at the expression on your answer sheet. What is the greatest whole number x can be?   6

20 How many two thousands are there, in two million?   1000

21 Look at the equation on your answer sheet: y equals five x, minus thirteen. What is y, when x is six?   17

22 From six squared, take five squared.   11

23 Find the mean of three, eleven, nine, and five.   7

24 Look at the expression on your answer sheet. Work out an approximate answer to this calculation.   (100)

25 Three coins are tossed. What is the probability that at least one lands on tails? $\frac{7}{8}$

1 The capacity of a tank has been measured to be sixty-two point one litres, to the nearest tenth of a litre. What is the minimum capacity the tank could have? — 62.05 litres
2 What is nought point eight seven, divided by one hundred? — 0.0087
3 What is minus seven, multiplied by plus two? — –14
4 Write nine tenths, as a decimal. — 0.9
5 Write nought point five five as a fraction, in its lowest terms. — $\frac{11}{20}$
6 What is one third of five sixths? — $\frac{5}{18}$
7 Write seven hundredths, as a percentage. — 7%
8 Write sixty hundredths, as a decimal. — 0.6
9 Divide one tenth by one third. — $\frac{3}{10}$
10 Divide nought point nought six by nought point one. — 0.6

11 A length of eighty-eight millimetres is to be reduced by one eighth. What is the reduced length? — 77 mm
12 Look at the equation on your answer sheet: a equals six c squared. What is a, when c is three? — 54
13 What is fifty, divided by five hundred? — 0.1
14 The area of a rectangle is thirty centimetres squared. The length of the rectangle is six centimetres. What is its perimeter? — 22 cm
15 Write, as a percentage, seven, divided by fifty. — 14
16 A train travels at an average speed of seventy kilometres per hour. A journey last three and a half hours. What is the length of the journey? — 245 km
17 Look at the expression on your answer sheet. What is the greatest whole number x can be? — 1
18 Multiply nought point two by eighty. — 16
19 How many four thousands are there, in four million? — 1000
20 Write down, as a decimal to one decimal place, an estimate for the value of fifty-one, divided by eighty-one. — (0.6)

21 Look at the equation on your answer sheet: seven equals forty-nine, minus seven x. What is x? — 6
22 Add five squared and three squared. — 34
23 Find the mean of three, five, four, five, eight, and five. — 5
24 Look at the expression on your answer sheet. Work out an approximate answer to this calculation. — (1)
25 How many nought point nought twos are there, in two hundred? — 10000

1 What is nought point nought eight, multiplied by nought point nought nought one? — 0.00008
2 Write one fifth, as a percentage. — 20%
3 Write eighty-one hundredths, as a decimal. — 0.81
4 Find the area of a square of side of length seven centimetres. — 49 cm²
5 Write ninety-nine per cent, as a decimal. — 0.99
6 The weight of a fishing line has been measured to be twenty-eight point four milligrams, to the nearest tenth of a milligram. What is the minimum weight the fishing line could be? — 28.35 mg
7 What is nought point nine three eight, multiplied by one thousand? — 938
8 Write five thousand ten-thousandths, as a decimal. — 0.5
9 Divide one eighth by one third. — $\frac{3}{8}$
10 What is nought point nought seven, divided by nought point nought nought nought seven? — 100

11 Sand and cement are mixed to make concrete in the ratio of four to one. What weight of sand is needed to mix with fifty kilograms of cement? — 200 kg
12 Multiply three hundred and eighty by thirty. — 11400
13 Decrease seven hundred by seventy per cent. — 210
14 Three angles of a quadrilateral are eighty degrees, one hundred and thirty degrees, and seventy degrees. What is the size of the fourth angle? — 80°
15 Look at the equation on your answer sheet. Find p, when q is six. — 2
16 There are eight hundred bolts in a container. How many bolts are there, in six hundred and thirty containers? — 504000
17 How many two thousands are there, in one million? — 500
18 A cyclist can maintain a constant speed of eight kilometres per hour, for two and a half hours. What distance can be covered in this time? — 20 km
19 Write down, as a decimal to one decimal place, an estimate for the value of eighty-three, divided by four hundred and nine. — (0.2)
20 Look at the expression on your answer sheet. What is the smallest whole number x can be? — 12

21 Estimate the value of sixty-nine per cent of five hundred and ninety-eight pounds. — £420
22 From four squared, subtract two cubed. — 8
23 Look at the equation on your answer sheet: seven x, minus fourteen, equals fourteen. What is x? — 4
24 Look at the expression on your answer sheet. Work out an approximate answer to this calculation. — (20)
25 A car is moving at a speed of fifty-four kilometres per hour. Write this speed in kilometres per minute. — 0.9 km/min

1 What is nought point nought nought three, multiplied by nought point nought two?   0.00006

2 Write twenty-one hundredths, as a percentage.   21%

3 What is six point six seven, divided by one hundred?   0.0667

4 Write thirty-six per cent as a fraction, in its lowest terms. $\frac{9}{25}$

5 What is minus five, multiplied by plus six?   −30

6 Write one quarter, as a decimal.   0.25

7 The length of a piece of wood has been measured as thirty-eight point seven centimetres, to the nearest millimetre. What is the maximum length the piece of wood could be?   38.75 cm

8 What is one fifth of three eighths? $\frac{3}{40}$

9 Write ninety thousandths, as a decimal.   0.09

10 What is nought point eight, divided by nought point nought two?   40

11 Two angles of a triangle are seventy-five degrees and ninety degrees. What is the size of the third angle?   15°

12 A length of one hundred and seventy metres is to be reduced by eighty per cent. What is the reduced length?   34

13 Look at the equation on your answer sheet: *d* equals three times the difference between *e* and four. Find *d*, when *e* is ten.   18

14 What is three thousand, three hundred, multiplied by fifty?   165000

15 Write, as a decimal, thirteen, divided by twenty-five.   0.52

16 What is nought point three, multiplied by three hundred?   90

17 Write down, as a whole number, an estimate for the value of six hundred and ninety-five, divided by three hundred and forty-nine.   (2)

18 A coach makes a journey of one hundred and seventy-five kilometres in two and a half hours. What is the average speed of the coach?   70 km/hr

19 Look at the expression on your answer sheet. What is the smallest whole number *x* can be?   13

20 How many six thousands are there, in three million?   500

21 Three dice are thrown together, and their scores are added. What is the probability of a total score of eighteen? $\frac{1}{216}$

22 Add nine squared to four squared.   97

23 Look at the equation on your answer sheet: *y* equals five *x*, minus fifteen. What is *y*, when *x* is eight?   25

24 Find the mean of six, five, two, and seven.   5

25 Look at the expression on your answer sheet. Work out an approximate answer to this calculation.   (100)

1 Find the area of a triangle with base of length six centimetres and height eight centimetres.   24 cm²

2 Write sixty-seven hundredths, as a percentage.   67%

3 What is nought point six nine four, multiplied by one hundred?   69.4

4 Write four fifths, as a decimal.   0.8

5 Write seventy-two per cent as a fraction, in its lowest terms. $\frac{18}{25}$

6 Write seventy ten-thousandths, as a decimal.   0.007

7 The capacity of a thimble has been measured as thirty-three point eight millilitres, to the nearest tenth of a millilitre. What is the minimum capacity, the thimble could have?   33.75 ml

8 What is one quarter of seven eighths? $\frac{7}{32}$

9 What is nought point nought six, divided by nought point nought nought one?   60

10 What is nought point two, multiplied by nought point two, multiplied by nought point nought one?   0.0004

11 The perimeter of a square is forty metres. What is the area of the square?   100 m²

12 An angle of one hundred and eighty degrees is to be decreased by five twelfths. What is the size of the reduced angle?   105°

13 Look at the equation on your answer sheet: *y* equals *x* squared, minus three. Find *y*, when *x* is seven.   46

14 What is eighteen, divided by nine thousand? Give your answer as a decimal.   0.002

15 Increase sixty-eight by two hundred per cent.   204

16 Estimate the value of eleven per cent of four hundred and fifty-one pounds.   (£49 or £45)

17 What is nought point nought nought five, multiplied by twenty?   0.1

18 Look at the expression on your answer sheet. What is the greatest whole number *x* can be?   8

19 How many twenty thousands are there, in one million?   50

20 A train travels at an average speed of ninety kilometres per hour for four and a half hours. How far is the journey?   405 km

21 The mean of five numbers is twenty-four. What is the sum of the five numbers?   120

22 Look at the equation on your answer sheet: *y* equals thirty-eight, minus six *x*. What is *y*, when *x* is four?   14

23 From six squared, take four squared.   20

24 How many nought point nought sevens are there, in one hundred and forty?   2000

25 Look at the expression on your answer sheet. Work out an approximate answer to this calculation.   (5)

1 The weight of a crate has been measured as three point nine nine tonnes, correct to two decimal places. What is the minimum weight, the crate could be?   3.985 tonne

2 Write seven tenths, as a decimal.   0.7

3 What is four point two nought one, multiplied by one thousand?   4201

4 Write seventeen hundredths, as a percentage.   17%

5 Write twenty-one per cent, as a decimal.   0.21

6 What is forty five, divided by minus nine?   −5

7 What is one half of three fifths? $\frac{3}{10}$

8 Write fifty thousandths, as a decimal.   0.05

9 What is nought point four, divided by nought point nought one?   40

10 What is one fifth, divided by one third? $\frac{3}{5}$

11 Three angles of a quadrilateral are one hundred and thirty degrees, seventy degrees, and eighty degrees. What is the size of the fourth angle?   80°

12 Multiply five thousand, three hundred by six hundred.   3180000

13 Write, as a decimal, eleven, divided by fifty.   0.22

14 Look at the equation on your answer sheet: p is four q squared. Find p, when q is four.   64

15 One hundred and twenty counters are divided into two piles in the ratio of five to seven. How many counters are there in the smallest pile?   50

16 How many two thousand, five hundreds are there, in one million?   400

17 Write down, as a whole number, an estimate for the value of two hundred and eight, divided by forty-four.   (5)

18 Multiply eight hundred by four hundred and eighty.   384000

19 Look at the expression on your answer sheet. What is the greatest whole number x can be?   17

20 How long does it take a lorry driver to complete two hundred and ten kilometres of a journey, at an average speed of ninety kilometres per hour?   2 hrs 20 min

21 Look at the equation on your answer sheet: nine x, minus fourteen, equals fifty-eight. What is x?   8

22 From seven squared, take five squared.   24

23 Look at the equation on your answer sheet: y equals eight x, minus eighteen. What is y, when x is seven?   38

24 How many nought point sixes are there, in one hundred and twenty?   200

25 Look at the expression on your answer sheet. Work out an approximate answer to this calculation.   (8000)

1 Find the area of a rectangle with length seven centimetres and width six centimetres.   42 cm²

2 Write one quarter, as a percentage.   25%

3 Write thirty-seven hundredths, as a decimal.   0.37

4 Divide seven point nought seven by one thousand.   0.00707

5 Write three per cent, as a decimal.   0.03

6 Write ninety ten-thousandths, as a decimal.   0.009

7 The weight of a shoe has been measured as nought point eight seven three kilograms, to the nearest gram. What is the maximum weight, the shoe could be?   0.8735 kg

8 What is nought point nought nought seven, multiplied by nought point nought one?   0.00007

9 What is half of five eighths? $\frac{5}{16}$

10 Divide nought point seven by nought point nought nought one.   700

11 A regiment of one hundred and twenty soldiers is to be increased in size by twenty per cent. How many soldiers will there then be?   144

12 The perimeter of a rectangle is eighteen millimetres. The length of the rectangle is seven millimetres. What is its area?   14 mm²

13 Look at the equation on your answer sheet: s is equal to thirty, minus five t. Find s, when t is six.   0

14 There are thirty felt-tip pens in a packet. How many pens will there be in eight hundred and thirty packets?   24900

15 Increase sixty-three litres of water by four ninths.   91 litres

16 Write down, as a decimal to one decimal place, an estimate for the value of forty-one, divided by fifty-eight.   (0.7)

17 What is nought point five, multiplied by seven hundred?   350

18 How many eight thousands are there, in four million?   500

19 Look at the expression on your answer sheet. What is the greatest whole number x can be?   3

20 A man cycles twenty-five kilometres in two and a half hours. What is his average speed?   10 km/hr

21 Look at the equation on your answer sheet: forty-two, minus four x, equals ten. What is x?   8

22 Find the mean of three, six, five, seven, four, and five.   5

23 From six squared, take away two cubed.   28

24 How many nought point nought fives are there in five hundred?   10000

25 Look at the expression on your answer sheet. Work out an approximate answer to this calculation.   (20)

1. The length of a rope has been measured as six point eight seven metres, to the nearest centimetre. What is the minimum length, the rope could be? — 6.865 m
2. Write twenty-nine hundredths, as a percentage. — 29%
3. Divide nought point six four by one hundred. — 0.0064
4. Write three quarters, as a decimal. — 0.75
5. Write nought point seven six as a fraction, in its lowest terms. — $\frac{19}{25}$
6. What is minus three, multiplied by minus two? — 6
7. What is half of one quarter? — $\frac{1}{8}$
8. Write ten hundredths, as a decimal. — 0.1
9. What is nought point nought three, divided by nought point one? — 0.3
10. What is nought point nought two, multiplied by nought point nought nought nought two? — 0.000004

11. The two angles of a triangle are both seventy degrees. What is the size of the third angle? — 40°
12. Multiply six hundred and seventy by three hundred. — 201000
13. Look at the equation on your answer sheet: d equals the square of the difference between e and three. Find d, when e is eight. — 25
14. Write, as a percentage, seven, divided by twenty-five. — 28%
15. A length of thirty-five metres is reduced by three sevenths. To what is the length reduced? — 20 m
16. Write down, as a decimal to one decimal place, an estimate for the value of sixty-two, divided by seventy-eight. — (0.8)
17. Multiply ninety by nought point nought seven. — 6.3
18. Look at the expression on your answer sheet. What is the smallest whole number x can be? — 7
19. How many four thousands are there, in two million? — 500
20. An object moves at a speed of thirty metres per second. How long does it take to go one hundred and thirty-five metres? — 4.5 sec

21. Estimate the value of eighty-nine per cent of three hundred and ninety-six pounds. — £360
22. Look at the equation on your answer sheet: y equals six x, minus thirteen. What is y, when x is eight? — 35
23. What is eight squared, added to six squared? — 100
24. How many nought point nought nines are there, in one hundred and eighty? — 2000
25. Look at the expression on your answer sheet. Work out an approximate answer to this calculation. — (1)

1. The distance between two towns has been measured as thirty-five point four kilometres, to the nearest tenth of a kilometre. What is the maximum the distance between the two towns could be? — 35.45 km
2. Write one fifth, as a decimal. — 0.2
3. Multiply seven point four seven by one thousand. — 7470
4. What is minus six, multiplied by plus four? — −24
5. Write nought point six nine, as a percentage. — 69%
6. Write fifty-nine hundredths, as a percentage. — 59%
7. Write eight thousand ten-thousandths, as a decimal. — 0.8
8. What is one third of one fifth? — $\frac{1}{15}$
9. What is nought point one, multiplied by nought point one, multiplied by nought point one? — 0.001
10. What is nought point nought five, divided by nought point one? — 0.5

11. In a car showroom, the ratio of used cars to new cars is five to three. There are thirty-six new cars. How many used cars are there? — 60
12. Look at the equation on your answer sheet: y equals x squared, minus two. Find y, when x is eight. — 62
13. Divide forty by eight hundred. — 0.05
14. A file costing four pounds has its price increased by nine per cent. What is the new price? — £4.36
15. The perimeter of a square is twelve metres. What is its area? — 9 m²
16. Multiply nine hundred and forty by seven hundred. — 658000
17. An object travels at a speed of thirty metres per second for three and a half seconds. How far has the object travelled? — 105 m
18. Write down, as an integer, an estimate for the value of five hundred and ninety-four, divided by twenty-nine. — (20)
19. Forty-one multiplied by thirty-eight equals one thousand, five hundred and fifty-eight. What is one thousand, five hundred and fifty-eight, divided by four point one? — 380
20. Look at the expression on your answer sheet. What is the greatest whole number x can be? — 4

21. Look at the equation on your answer sheet: five x, minus twelve, equals thirty-three. What is x? — 9
22. What is three squared, taken from eight squared? — 55
23. How many nought point threes are there, in one hundred and twenty? — 400
24. Look at the equation on your answer sheet: y equals six x, minus twelve. What is y, when x is eight? — 36
25. Look at the expression on your answer sheet. Work out an approximate answer to this calculation. — (100)

E21

1 Find the area of a rectangle of length nine centimetres and width seven centimetres.   63 cm²

2 Write seventy-seven hundredths, as a percentage.   77%

3 Divide eighty-seven point six by one thousand.   0.0876

4 Write three tenths, as a decimal.   0.3

5 Write nought point six four as a fraction, in its lowest terms. $\frac{16}{25}$

6 Write thirty thousandths, as a decimal.   0.03

7 A scuba diver makes a dive to a depth of eighty point three metres. His gauge gives a measurement, to the nearest tenth of a metre. What could be the maximum depth reached?   80.35 m

8 Multiply nought point nought three by nought point nought two.   0.0006

9 What is one quarter of five eighths? $\frac{5}{32}$

10 Divide nought point eight by nought point nought one.   80

11 Three angles of a quadrilateral are ninety degrees, sixty degrees, and one hundred and forty degrees. What is the size of the fourth angle?   70°

12 Increase fifty-five by sixty per cent.   88

13 In a packet, there are notes to the value of two hundred dollars. What is the total value of five hundred and seventy packets? $114000

14 Look at the equation on your answer sheet: y equals thirteen, minus two x. Find y, when x is six.   1

15 Write, as a decimal, eleven, divided by twenty-five.   0.44

16 Look at the expression on your answer sheet. What is the greatest whole number x can be?   8

17 What is nought point nought nought seven, multiplied by sixty?   0.42

18 How many four thousands are there, in one million?   250

19 Estimate the value of sixty-two per cent of nine hundred and four pounds.   (£540 to £560)

20 A man drives one hundred and fifty kilometres in two and a half hours. What is his average speed for the journey?   60 km/hr

21 A van is moving at a speed of nought point seven kilometres per minute. Write this speed in kilometres per hour.   42 km/hr

22 The mean of six numbers is twelve. What is the sum of the six numbers?   72

23 Look at the equation on your answer sheet: eight x, minus fourteen, equals thirty-four. What is x?   6

24 What is seven squared, added to two squared?   53

25 Look at the expression on your answer sheet. Work out an approximate answer to this calculation.   (0.5)

E22

1 A temperature has been measured as thirty point nine degrees Centigrade, to one decimal place. What is the minimum the temperature could be?   30.85°C

2 What is twelve, divided by minus three?   −4

3 Write seven tenths, as a percentage.   70%

4 Write fifty-seven hundredths, as a decimal.   0.57

5 Write nought point nought four, as a percentage.   4%

6 What is half of three quarters? $\frac{3}{8}$

7 Multiply nine point two seven by one hundred.   927

8 Write twenty ten-thousandths, as a decimal.   0.002

9 What is nought point five, divided by nought point nought nought one?   500

10 What is one seventh, divided by one quarter? $\frac{4}{7}$

11 A tank contains eighty litres of oil. This capacity is to be increased by five eighths. How many litres of oil will there be?   130 litres

12 What is sixty, divided by six thousand?   0.01

13 Look at the equation on your answer sheet: C equals two times the answer to d add three. Find C, when d is nine.   24

14 Write, as a percentage, twenty-one, divided by twenty-five.   84%

15 The area of a rectangle is twenty-eight metres squared. The width of the rectangle is four metres. What is its perimeter?   22 m

16 What is eight hundred and eighty, multiplied by seven hundred?   616000

17 Write down, as a decimal to one decimal place, an estimate for the value of forty-two, divided by seventy-seven.   (0.5)

18 How many five thousands are there, in four million?   800

19 Look at the expression on your answer sheet. What is the smallest whole number x can be?   11

20 A lorry is driven at an average speed of ninety kilometres per hour for two hours and forty minutes. How far has the lorry travelled in this time?   240 km

21 Look at the equation on your answer sheet: y equals thirty-eight, minus five x. What is y, when x is seven?   3

22 From three cubed, subtract two squared.   23

23 Find the mean of five, seven, two, and ten.   6

24 How many nought point nought sixes are there, in two hundred and forty?   4000

25 Look at the expression on your answer sheet. Work out an approximate answer to this calculation.   (400)

E23

Answers

1. Find the area of a triangle with base of length four centimetres and height eight centimetres. **16 cm²**

2. What is sixty-seven point four, divided by one hundred? **0.674**

3. Write two fifths, as a decimal. **0.4**

4. Write thirty-one hundredths, as a percentage. **31%**

5. Write thirty-seven per cent, as a decimal. **0.37**

6. The length of a wire has been measured as eight point four nought metres, to the nearest centimetre. What is the minimum length the wire could be? **8.395 m**

7. What is nought point nought nought four, multiplied by nought point nought two? **0.00008**

8. What is two sevenths, divided by one third? **$\frac{6}{7}$**

9. Write four thousand ten-thousandths, as a decimal. **0.4**

10. What is nought point seven, divided by nought point nought one? **70**

11. An amount of one hundred and sixty pounds is shared in the ratio of five to three. What is the larger share? **£100**

12. What is seven thousand, three hundred, multiplied by thirty? **219000**

13. Look at the equation on your answer sheet: p equals the square of the sum of q and two. Find p, when q is four. **36**

14. The two angles of a triangle are seventy-five degrees and eighty-five degrees. What is the size of the third angle of the triangle? **20°**

15. Decrease a weight of fifty kilograms by seven tenths. **15 kg**

16. Look at the expression on your answer sheet. What is the greatest whole number x can be? **4**

17. What is nought point nought nought two, multiplied by twenty? **0.04**

18. Write down, as a decimal to two decimal places, an estimate for the value of two hundred and four, divided by eight hundred and four. **(0.25)**

19. A girl cycles four kilometres in forty minutes. What is her average speed for the journey? **6 km/hr**

20. How many fifteen hundreds are there, in three million? **2000**

21. Look at the equation on your answer sheet: five x, plus nine, equals twenty-four. What is x? **3**

22. Find the mean of five, six, two, four, and three. **4**

23. From six squared, subtract three squared. **27**

24. How many nought point nought fives are there, in two hundred and fifty? **5000**

25. Look at the expression on your answer sheet. Work out an approximate answer to this calculation. **(5)**

E24

Answers

1. The capacity of a teaspoon has been measured to be eight point five millilitres, to the nearest tenth of a millilitre. What is the minimum capacity, the teaspoon could have? **8.45 ml**

2. Write four fifths, as a percentage. **80%**

3. Find the area of a square of side of length six centimetres. **36 cm²**

4. Write sixty-nine hundredths, as a decimal. **0.69**

5. Write forty-eight per cent as a fraction, in its lowest terms. **$\frac{12}{25}$**

6. What is fifty two point three, divided by one hundred? **0.523**

7. Write six thousand ten-thousandths, as a decimal. **0.6**

8. What is one half of five sixths? **$\frac{5}{12}$**

9. What is nought point two, divided by nought point nought nought one? **200**

10. What is nought point nought four, multiplied by nought point nought one? **0.0004**

11. The perimeter of a square is twenty-eight millimetres. What is the area of the square? **49 mm²**

12. Increase twenty-two by fifty per cent. **33**

13. In a stadium, there are forty areas, each of which has a maximum capacity of three hundred spectators. What is the total capacity of the stadium? **12000**

14. Look at the equation on your answer sheet: a equals two c squared. Find a, when c is eight. **128**

15. Increase eighteen by one ninth. **20**

16. Write down, as an integer, an estimate for the value of fifty-seven, divided by thirty-two. **(2)**

17. What is nought point nought three, multiplied by five hundred? **15**

18. An object travels at a speed of thirty kilometres per hour. How long does it take to travel eighty kilometres? **2 hrs 40 min**

19. Look at the expression on your answer sheet. What is the greatest whole number x could be? **9**

20. How many forty thousands are there, in one million? **25**

21. How many nought point nought threes are there, in one hundred and fifty? **5000**

22. Take six squared from eight squared. **28**

23. Look at the equation on your answer sheet: y equals seven x, minus thirteen. What is y, when x is six? **29**

24. Estimate the value of forty-nine per cent of one hundred and ninety-five. **(100)**

25. Look at the expression on your answer sheet. Work out an approximate answer to this calculation. **(50)**

E25
Answers

1. A wind speed has been measured as twelve point four metres per second, to one decimal place. What is the greatest speed, the wind speed could be? 12.45 m/s
2. What is fifty-six point nine, divided by one thousand? 0.0569
3. Write three fifths, as a decimal. 0.6
4. Write nought point two eight as a fraction, in its lowest terms. $\frac{7}{25}$
5. What is minus fifteen, divided by minus three? 5
6. What is nought point nought nought two, multiplied by nought point nought three? 0.00006
7. Write nine hundredths, as a percentage. 9%
8. What is one third of one sixth? $\frac{1}{18}$
9. Write fifty hundredths, as a decimal. 0.5
10. What is nought point nought nine, divided by nought point nought nought nine? 10

11. Three angles of a quadrilateral are one hundred degrees, seventy degrees, and one hundred degrees. What is the size of the fourth angle? 90°
12. Look at the equation on your answer sheet: y equals x squared, minus 4. Find y, when x is eleven. 117
13. Write, as a percentage, eleven, divided by fifty. 22%
14. Divide seventy by seven hundred. Give your answer as a decimal. 0.1
15. What is six hundred and eighty, multiplied by twenty? 13600
16. A train travels at an average speed of sixty kilometres per hour for three and a half hours. What distance has been travelled in this time? 210 km
17. Write down, as a decimal to one decimal place, an estimate for seventy-seven divided by two hundred and four. (0.4)
18. Look at the expression on your answer sheet. What is the greatest whole number x can be? 10
19. What is nought point nought nought four, multiplied by eighty? 0.32
20. Fifty-nine multiplied by nineteen equals one thousand, one hundred and twenty-one. What is one thousand, one hundred and twenty-one, divided by nought point one nine? 5900

21. Look at the expression on your answer sheet. Work out an approximate answer to this calculation. (10)
22. Look at the equation on your answer sheet: six x, minus twelve, equals thirty. What is x? 7
23. Add eight squared to three squared. 73
24. Look at the equation on your answer sheet: y equals five x, minus thirteen. What is y, when x is eight? 27
25. An object is moving at a speed of ten metres per second. Write this speed in kilometres per hour. 36 km/hr

E26
Answers

1. An angle has been measured as thirty point four degrees, to one decimal place. What is the minimum size, the angle could be? 30.35°
2. Write forty-one hundredths, as a percentage. 41%
3. What is minus four, multiplied by minus five? 20
4. Write four fifths, as a decimal. 0.8
5. Write eighty-eight per cent as a fraction, in its lowest terms. $\frac{22}{25}$
6. What is nought point nought nought one, multiplied by nought point nought one? 0.00001
7. What is three eighths, divided by one third? $\frac{9}{8}$ or $1\frac{1}{8}$
8. What is eight point seven six, multiplied by one hundred? 876
9. Write fifty ten-thousandths, as a decimal. 0.005
10. What is nought point four, divided by nought point nought two? 20

11. The three angles of a quadrilateral are one hundred degrees, seventy degrees, and one hundred degrees. What size is the fourth angle? 90°
12. In a garden, the ratio of blue flowers to white flowers is six to thirteen. There are forty-two blue flowers. How many white flowers are there? 91
13. Increase eighty-eight grams by three eighths. 121 g
14. Multiply six thousand, one hundred by eighty. 488000
15. Look at the equation on your answer sheet: a equals twenty-four, divided by the sum of c and three. Find a, when c is zero. 8
16. What is nought point nine, multiplied by four hundred? 360
17. What is five hundred and ninety, multiplied by sixty? 35400
18. Write down, as a decimal to one decimal place, an estimate for the value of thirty-eight, divided by seventy-nine. (0.5)
19. Look at the expression on your answer sheet. What is the smallest whole number x can be? 76
20. A boy walks at a speed of six kilometres per hour. How long does he take to walk twenty-one kilometres? 3 hrs 30 min

21. Look at the expression on your answer sheet. Work out an approximate answer to this calculation. (3)
22. Find the mean of two, nine, three, eight, five, and three. 5
23. Look at the equation on your answer sheet: forty, minus six x, equals ten. What is x? 5
24. Add nine squared and eight squared. 145
25. How many nought point nought fives are there, in one hundred? 2000

E27

1 Find the area of a triangle with base of length six metres and height of four metres. 12 cm²
2 Write sixty-one hundredths, as a percentage. 61%
3 What is sixty-eight point six, divided by one thousand? 0.0686
4 Write one tenth, as a decimal. 0.1
5 Write nought point six four as a fraction, in its lowest terms. $\frac{16}{25}$
6 What is nought point nought three, multiplied by nought point three? 0.009
7 The capacity of a jug has been measured as nought point four two litres, to the nearest ten millilitres.
What is the minimum capacity the jug can have? 0.415 litres
8 What is one half of one eighth? $\frac{1}{16}$
9 Write ninety hundredths, as a decimal. 0.9
10 What is nought point nought seven, divided by nought point nought one? 7

11 Three are four hundred and sixty people on a train. This number is reduced by ten per cent at the next station.
How many people remain on the train? 414
12 Look at the equation on your answer sheet: *d* equals four times the answer to *e*, take away two.
Find *d*, when *e* is nine. 28
13 What is six thousand, three hundred, multiplied by forty? 252000
14 Reduce thirty-five by two sevenths. 25
15 The perimeter of a square is eight centimetres. What is its area? 4 cm²
16 How many ten thousands are there in one million? 100
17 Estimate the value of twenty-two per cent of three hundred and four. (66 or 60)
18 What is nought point five, multiplied by five thousand? 2500
19 Look at the expression on your answer sheet. What is the greatest whole number *x* can be? 21
20 A man drives three hundred and fifteen kilometres in three and a half hours.
What is his average speed for the journey? 90 km/hr

21 The mean of five numbers is thirty-three. What is the sum of the five numbers? 165
22 From seven squared, take two cubed. 41
23 Look at the equation on your answer sheet: *y* equals eight *x*, minus twelve.
What is *y*, when *x* is seven? 44
24 Two dice are rolled, and their scores added. What is the probability of a total score of less than five? $\frac{1}{6}$
25 Look at the expression on your answer sheet. Work out an approximate answer to this calculation. (10)

E28

1 The length of a piece of fabric has been measured as two point three nought metres, to the nearest centimetre.
What is the minimum length, the piece of fabric could be? 2.295 m
2 Write nine tenths, as a decimal. 0.9
3 What is nought point two eight four, multiplied by one thousand? 284
4 Write eleven per cent, as a decimal. 0.11
5 What is minus eight, multiplied by minus two? 16
6 Write ninety-seven hundredths, as a percentage. 97%
7 What is nought point nought one, multiplied by nought point nought one, multiplied by nought point two? 0.00002
8 What is one third of one eighth? $\frac{1}{24}$
9 Write sixty ten-thousandths, as a decimal. 0.006
10 What is nought point two, divided by nought point nought one? 20

11 The number of tickets available through a booking agent is to be increased by two hundred per cent from thirty-five tickets. How many tickets will now be available? 105
12 Multiply three thousand, five hundred by thirty. 105000
13 Look at the equation on your answer sheet: *y* is six *x* squared.
Find *y*, when *x* is five. 150
14 Two angles of a triangle are sixty degrees and seventy degrees.
What is the size of the third angle? 50°
15 Reduce forty-five by five ninths. 20
16 What is nought point nought nought two, multiplied by four hundred? 0.8
17 How many thousands are there, in two million? 2000
18 Write down, as a whole number, an estimate for the value of two hundred and ninety-seven, divided by thirty-one. (10)
19 A van is driven at an average speed of eighty kilometres per hour for three and a half hours.
How many kilometres would have been travelled in this time? 280 km
20 Look at the expression on your answer sheet. What is the minimum whole number *x* could be? 1

21 Look at the equation on your answer sheet: seven *x*, minus thirteen, equals fifty. What is *x*? 9
22 Find the mean of six, one, four, five, and four. 4
23 Add six squared to four squared. 52
24 How many nought point nought fives are there, in one thousand? 20000
25 Look at the expression on your answer sheet. Work out an approximate answer to this calculation. (100)

E29

1 Find the area of a rectangle of length eight metres and width four metres. 32 cm²
2 Write one fifth, as a percentage. 20%
3 Write forty-seven hundredths, as a decimal. 0.47
4 What is five point five one, multiplied by one hundred? 551
5 Write nought point two four as a fraction, in its lowest terms. $\frac{6}{25}$
6 What is one quarter of one fifth? $\frac{1}{20}$
7 A temperature has been measured as forty-four point three degrees Fahrenheit, to one decimal place. What is the maximum, the temperature could be? 44.35°F
8 Write eight hundred thousandths, as a decimal. 0.8
9 What is nought point nought nine, divided by nought point nought one? 9
10 What is one third, divided by one half? $\frac{2}{3}$

11 The perimeter of a rectangle is twenty-six millimetres. The length of the rectangle is seven millimetres. What is its area? 42 mm²
12 Write, as a decimal, three, divided by twenty-five. 0.12
13 Divide sixty by one thousand, two hundred. 0.05
14 Look at the equation on your answer sheet: d equals the square of the sum of e and three. Find d, when e is six. 81
15 A block of cheese of weight two hundred and fifty grams is divided in the ratio of twelve to thirteen. What is the weight of the smaller piece? 120 g
16 Multiply seven hundred and thirty by ninety. 65700
17 Look at the expression on your answer sheet. What is the greatest whole number x can be? 6
18 How many three thousands are there, in three million? 1000
19 Write down, as a decimal to one decimal place, an estimate for the value of twenty-five, divided by eighty-two. (0.3)
20 A man drives one hundred and thirty-five kilometres in four and a half hours. What is his average speed for the journey? 30 km/hr

21 Look at the expression on your answer sheet. Work out an approximate answer to this calculation. (6)
22 Estimate the value of sixty-nine per cent of five hundred and ninety-eight pounds. £420
23 Look at the equation on your answer sheet: y equals seven x, minus twelve. What is y, when x is six? 30
24 From three squared, take two squared. 5
25 A plane is moving at a speed of three hundred and sixty kilometres per hour. Write this speed in metres per second. 100 m/s

E30

1 The length of a journey has been measured as eighteen point three two kilometres, to four significant figures. What is the minimum distance, the length of the journey could be? 18.315 km
2 Write seventy-one hundredths, as a percentage. 71%
3 What is nought point seven three, multiplied by one thousand? 730
4 Write seven tenths, as a decimal. 0.7
5 Write nought point nought six, as a percentage. 6%
6 What is nought point nought nought four, multiplied by nought point nought one? 0.00004
7 What is minus eight, divided by plus two? −4
8 What is half of three eighths? $\frac{3}{16}$
9 Write nine hundred thousandths, as a decimal. 0.9
10 What is nought point nought two, divided by nought point nought nought one? 20

11 The three angles of a quadrilateral are one hundred and ten degrees, one hundred and thirty degrees, and seventy degrees. What is the size of the fourth angle? 50°
12 Look at the equation on your answer sheet: y equals x squared, plus three. Find y, when x is seven. 52
13 Increase thirty-six by five sixths. 66
14 Write, as a percentage, seventeen, divided by twenty-five. 68%
15 Multiply five hundred and eighty by twenty. 11600
16 Look at the expression on your answer sheet. What is the greatest whole number x can be? 8
17 A girl can cycle at an average speed of twelve kilometres per hour. How long will it take her to cycle twenty-eight kilometres? 2 hrs 20 min
18 What is nought point nought nought nine, multiplied by seventy? 0.63
19 Sixteen multiplied by thirty-six is five hundred and seventy-six. What is five hundred and seventy-six, divided by nought point three six? 1600
20 Write down, as a decimal to one decimal place, an estimate for the value of seven hundred and forty-nine, divided by eight hundred and twenty-one. (0.9)

21 Look at the expression on your answer sheet. Work out an approximate answer to this calculation. (400)
22 Add two squared to three cubed. 31
23 Look at the equation on your answer sheet: six x, minus eleven, equals thirty-seven. What is x? 8
24 How many nought point nought sixes are there, in one hundred and twenty? 2000
25 The mean of six numbers is fifteen. What is the sum of the six numbers? 90

1 Find the area of a triangle of base of length six metres and height nine metres. — 27 m²

2 Divide seven point nought seven by one hundred. — 0.0707

3 Write eighty-seven hundredths, as a decimal. — 0.87

4 Write three quarters, as a percentage. — 75%

5 Write eight per cent, as a decimal. — 0.08

6 An angle has been measured as fourteen point one degrees, to one decimal place. What is the minimum size the angle could be? — 14.05°

7 What is nought point three, multiplied by nought point nought nought two? — 0.0006

8 Write twenty thousandths, as a decimal. — 0.02

9 What is nought point eight, divided by nought point nought one? — 80

10 What is one fifth of one eighth? — $\frac{1}{40}$

11 An eighty-pound car repair bill has been increased by forty per cent by some additional work. What will be the amount of the new bill? — £112

12 The perimeter of a square is sixteen millimetres. What is its area? — 16 mm²

13 Multiply three hundred and sixty by two hundred. — 72000

14 Look at the equation on your answer sheet: C equals twenty-five, minus three x. Find C, when x is six. — 7

15 Increase ninety-nine by five elevenths. — 144

16 An object travels at a speed of thirty metres per second for two and a half seconds. How far has it travelled in this time? — 75 m

17 Multiply nought point nought six by six hundred. — 36

18 Estimate the value of nine per cent of seven hundred and forty-eight pounds. — (£75 or £68 or £80)

19 Look at the expression on your answer sheet. What is the greatest whole number x can be? — 23

20 How many forty thousands are there, in one million? — 25

21 Look at the expression on your answer sheet. Work out an approximate answer to this calculation. — (3)

22 Look at the equation on your answer sheet: y equals forty-five, minus four x. What is y, when x is six? — 21

23 Look at the equation on your answer sheet: five x, minus sixteen, equals nineteen. What is x? — 7

24 Add six squared to seven squared. — 85

25 How many nought point fives are there, in one thousand? — 2000

1 An electrical current has been measured as five point four two amps, to three significant figures. What is the maximum number of amps, the current could be? — 5.425 amps

2 Write three quarters, as a decimal. — 0.75

3 Write eighty-nine hundredths, as a percentage. — 89%

4 Find the area of a square of side of length eight centimetres. — 64 cm²

5 Write eighty-four per cent as a fraction, in its lowest terms. — $\frac{21}{25}$

6 What is one fifth of nine tenths? — $\frac{9}{50}$

7 Multiply nought point eight five one by one hundred. — 85.1

8 What is nought point one, multiplied by nought point nought one, multiplied by nought point nought three? — 0.00003

9 Write seventy hundredths, as a decimal. — 0.7

10 What is nought point nought eight, divided by nought point nought nought eight? — 10

11 There are two thousand, three hundred people in a cinema. Five per cent leave early. How many then remain in the cinema? — 2185

12 Look at the equation on your answer sheet: p equals the square of the sum of q and three. Find p, when q is six. — 81

13 Divide seventy by seven thousand. — 0.01

14 The two angles of a triangle are sixty-five degrees and seventy-five degrees. What size is the third angle? — 40°

15 In a survey, the ratio of coffee drinkers to tea drinkers is ten to seven. Forty-nine people were tea drinkers. How many coffee drinkers were there? — 70

16 A man walks four kilometres in forty minutes. What was his average speed? — 6 km/hr

17 Write down, as a whole number, an estimate for the value of seven hundred and nine, divided by ninety-six. — (7)

18 What is nought point nought nought eight, multiplied by eighty? — 0.64

19 Look at the expression on your answer sheet. What is the smallest whole number x can be? — 5

20 How many twenty-five thousands are there, in one million? — 40

21 Look at the expression on your answer sheet. Work out an approximate answer to this calculation. — (200)

22 Estimate the value of eighty-two per cent of eight hundred and two. — (640 or 650)

23 Look at the equation on your answer sheet: y equals eight x, minus twelve. What is y, when x is seven? — 44

24 Subtract three squared from four squared. — 7

25 A pool is being emptied at the rate of eight litres per second. Write this rate in litres per minute. — 480 litres/min

1 A temperature has been recorded as minus four point three degrees Centigrade, to one decimal place. What is the minimum, the temperature could be? −4.35°C
2 What is twenty, divided by minus five? −4
3 Divide four point six by one thousand. 0.0046
4 Write four fifths, as a decimal. 0.8
5 Write nought point one eight as a fraction, in its lowest terms. $\frac{9}{50}$
6 Write twenty-seven hundredths, as a percentage. 27%
7 What is nought point nought three, multiplied by nought point nought nought two? 0.00006
8 What is one half of one fifth? $\frac{1}{10}$
9 Write nine thousand ten-thousandths, as a decimal. 0.9
10 What is nought point eight, divided by nought point nought four? 20

11 Five eighths of a sixteen-dollar budget has been spent. How much is left? $6
12 Look at the equation on your answer sheet: y equals three times the answer to x, take away two. Find y, when x is eleven. 27
13 Multiply eight hundred and twenty by forty. 32800
14 What is, as a percentage, nine, divided by twenty-five? 36%
15 The area of a rectangle is ten centimetres squared. The length of the rectangle is five centimetres. What is its perimeter? 14 cm
16 Write down, as an integer, an estimate for the value of six hundred and thirty-five, divided by thirty-two. (20)
17 What is nine hundred and forty, multiplied by sixty? 56400
18 How many three thousands are there, in three million? 1000
19 Look at the expression on your answer sheet. What is the greatest whole number x can be? 7
20 A van is driven at a steady speed of sixty kilometres per hour for three hours and forty-five minutes. How far has the van travelled? 225 km

21 Look at the expression on your answer sheet. Work out an approximate answer to this calculation. (4)
22 Find the mean of five, three, nine, and three. 5
23 Look at the equation on your answer sheet: six x, minus thirteen, equals eleven. What is x? 4
24 How many nought point sixes are there, in six hundred? 1000
25 Add seven squared to four squared. 65

1 A period of time is measured as five point nine seconds, to two significant figures. What is the minimum, the time could be? 5.85 sec
2 Write three fifths, as a percentage. 60%
3 Write ninety-three hundredths, as a decimal. 0.93
4 Find the area of a rectangle of length eight centimetres and width four centimetres. 32 cm²
5 Write six per cent, as a decimal. 0.06
6 What is nought point nought two, multiplied by nought point three? 0.006
7 Divide four point six five by one hundred. 0.0465
8 What is nought point six, divided by nought point nought nought one? 600
9 What is one third of seven eighths? $\frac{7}{24}$
10 Write seventy thousandths, as a decimal. 0.07

11 The perimeter of a square is twenty-four metres. What is its area? 36 m²
12 Write, as a percentage, three, divided by fifty. 6%
13 Multiply four hundred and fifty by thirty. 13500
14 Decrease eighteen metres by four ninths. 10 m
15 Look at the equation on your answer sheet: y equals twenty, divided by the sum of x and two. Find y, when x is eight. 2
16 How many five thousands are there, in four million? 800
17 What is nought point nought nought four, multiplied by seven hundred? 2.8
18 A car is driven at an average speed of eighty kilometres per hour. How long does it take to travel three hundred kilometres? 3 hrs 45 min
19 Estimate the value of eighteen per cent of six hundred and ninety-seven. (140)
20 Look at the expression on your answer sheet. What is the smallest whole number x can be? 19

21 Look at the equation on your answer sheet: y equals forty-two, minus four x. What is y, when x is nine? 6
22 Add eight squared and two cubed. 72
23 Look at the equation on your answer sheet: seven x, minus twelve, equals fifty-one. What is x? 9
24 Look at the expression on your answer sheet. Work out an approximate answer to this calculation. (400)
25 How many nought point nought twos are there, in two thousand? 100000

E35

1 An angle has been measured as twenty-three point two degrees, to one decimal place. What is the maximum size, this angle could be? — 23.25°
2 What is minus four, divided by plus two? — −2
3 Multiply nought point eight eight four by one hundred. — 88.4
4 Write ninety-one hundredths, as a decimal. — 0.91
5 Write nought point nought one, as a percentage. — 1%
6 What is one third of one third? — $\frac{1}{9}$
7 Multiply nought point nought two by nought point nought two. — 0.0004
8 Write two hundred thousandths, as a decimal. — 0.2
9 Write one quarter, as a percentage. — 25%
10 Divide nought point six by nought point nought nought six. — 100

11 Three angles of a quadrilateral are one hundred degrees, one hundred and ten degrees, and ninety degrees. What is the size of the fourth angle? — 60°
12 The sum of ninety-six pounds is divided between Louise and Michael in the ratio five to three. How much did Michael receive? — £36
13 Look at the equation on your answer sheet: d equals three e squared. Find d, when e equals seven. — 147
14 What is three hundred and ninety, multiplied by forty? — 15600
15 The price of a ticket on a ferry has increased by twenty-five per cent from eighty pounds. What is the new price? — £100
16 What is nought point nine, multiplied by nine thousand? — 8100
17 Look at the expression on your answer sheet. What is the smallest whole number x can be? — 6
18 A driver takes forty minutes to travel twenty kilometres. What is the average speed for the journey? — 30 km/hr
19 Write down, as a decimal to one decimal place, an estimate for two hundred and ninety-nine, divided by eight hundred and eight. — (0.4)
20 Thirty-six multiplied by forty-nine is one thousand, seven hundred and sixty-four. What is one thousand, seven hundred and sixty-four, divided by nought point three six? — 4900

21 Look at the expression on your answer sheet. Work out an approximate answer to this calculation. — (4000)
22 Look at the equation on your answer sheet: y equals eight x, minus fifteen. What is y, when x is seven? — 41
23 Add eight squared to seven squared. — 113
24 Divide one hundred and fifty by nought point nought five. — 3000
25 The mean of five numbers is fourteen. What is the sum of the five numbers? — 70

E36

1 Find the area of a triangle of base of length eight centimetres, and height seven centimetres. — 28 cm²
2 Write seven tenths, as a decimal. — 0.7
3 Multiply nought point eight one by one thousand. — 810
4 Write sixty-three hundredths, as a percentage. — 63%
5 Write nought point four as a fraction, in its lowest terms. — $\frac{2}{5}$
6 The length of a rope has been measured as twelve point nought centimetres, to the nearest millimetre. What is the minimum length, the rope could be? — 11.95 cm
7 Multiply nought point nought nought one by nought point nought nought one. — 0.000001
8 What is one quarter of five sixths? — $\frac{5}{24}$
9 Write eighty ten-thousandths, as a decimal. — 0.008
10 What is nought point nought four, divided by nought point nought nought four? — 10

11 The area of a rectangle is thirty-two millimetres squared. The width of the rectangle is four millimetres. What is the length of the perimeter? — 24 mm
12 Decrease five hundred by one per cent. — 495
13 Look at the equation on your answer sheet: y equals x squared, plus four. Find y, when x is six. — 40
14 Write, as a decimal, eleven, divided by twenty-five. — 0.44
15 Divide sixty by six hundred. Give your answer as a decimal. — 0.1
16 What is four hundred and seventy, multiplied by ninety? — 42300
17 Write down, as a whole number, an estimate for the value of four hundred and seventy-four, divided by fifty-three. — (9)
18 What is nought point nought seven, multiplied by fifty? — 3.5
19 Look at the expression on your answer sheet. What is the smallest whole number x could be? — 14
20 A train is travelling at an average speed of ninety kilometres per hour. How far will the train travel in three hours twenty minutes? — 300 km

21 Look at the expression on your answer sheet. Work out an approximate answer to this calculation. — (4)
22 Find the mean of seven, four, two, five, and two. — 4
23 Look at the equation on your answer sheet: forty-one, minus five x, equals eleven. What is x? — 6
24 How many nought point nought threes are there, in one hundred and twenty? — 4000
25 Three coins are tossed. What is the probability that at least one lands on heads? — $\frac{7}{8}$

1 A temperature has been measured as ten point one degrees Fahrenheit, to one tenth of a degree. What is the minimum, the temperature could be? **10.05°F**

2 Write ninety-nine hundredths, as a percentage. **99%**

3 Write two fifths, as a decimal. **0.4**

4 What is minus fifteen, divided by minus three? **5**

5 Write nought point nought nine, as a percentage. **9%**

6 Find one fifth of one tenth. $\frac{1}{50}$

7 Divide thirty-two point three by one thousand. **0.0323**

8 What is nought point nought four, divided by nought point one? **0.4**

9 Write thirty hundredths, as a decimal. **0.3**

10 What is nought point nought three, multiplied by nought point nought nought two? **0.00006**

11 In a class, the ratio of boys to girls is one to two. There are nine boys. How many girls are there? **18**

12 Look at the equation on your answer sheet: p equals twenty, minus four q. Find p, when q is four. **4**

13 Multiply eight thousand, four hundred by two hundred. **1680000**

14 Two angles of a triangle are both sixty degrees. What is the size of the third angle? **60°**

15 Increase forty-five dollars by one ninth. **$50**

16 Write down, as a whole number, an estimate for the value of seven hundred and nineteen, divided by fifty-two. **(14)**

17 A motorbike travels sixty kilometres in just forty minutes. What is the average speed of the motorbike? **90 km/hr**

18 What is nought point nought nought three, multiplied by two hundred? **0.6**

19 Look at the expression on your answer sheet. What is the greatest whole number x could be? **11**

20 How many two thousands are there, in one million? **500**

21 Look at the equation on your answer sheet: y equals five x, minus twelve. What is y, when x is six? **18**

22 Add eight squared to nine squared. **145**

23 Estimate the value of fifty-eight per cent of seven hundred and ninety-nine. **(480)**

24 Look at the expression on your answer sheet. Work out an approximate answer to this calculation. **(200)**

25 How many nought point nought twos are there, in one hundred? **5000**

1 The weight of a container is measured as nought point four two four kilograms, to the nearest gram. What is the maximum weight, the container could be? **0.4245 kg**

2 Write nine tenths, as a percentage. **90%**

3 Divide eleven point eight by one hundred. **0.118**

4 What is minus two, multiplied by minus two? **4**

5 Write five per cent, as a decimal. **0.05**

6 What is one seventh, divided by one half? $\frac{2}{7}$

7 What is nought point one, multiplied by nought point nought two, multiplied by nought point nought nought three? **0.000006**

8 Write forty-three hundredths, as a decimal. **0.43**

9 Write forty thousandths, as a decimal. **0.04**

10 What is nought point nought two, multiplied by nought point nought nought three? **0.00006**

11 There are three hundred people who work at a factory. The workforce is to be increased by sixty per cent. What size will the workforce then be? **480**

12 Multiply three hundred and seventy by twenty. **7400**

13 Look at the equation on your answer sheet: d equals twice the answer to e add three. Find d, when e is fourteen. **34**

14 The perimeter of a square is four centimetres. What is its area? **1 cm²**

15 Reduce fifty litres by nine tenths. **5 litres**

16 Multiply six hundred and thirty by eighty. **50400**

17 A cyclist can maintain a speed of twelve kilometres per hour. How long does it take to cycle nine kilometres? **0 hrs 45 min**

18 How many five thousands are there, in forty million? **8000**

19 Write down, as a decimal to one decimal place, an estimate for the value of forty-three, divided by eighty-three. **(0.5)**

20 Look at the expression on your answer sheet. What is the smallest whole number x could be? **4**

21 Look at the expression on your answer sheet. Work out an approximate answer to this calculation. **(500)**

22 Look at the equation on your answer sheet: y equals eight x, minus fourteen. What is y, when x is six? **34**

23 Look at the equation on your answer sheet: five x, minus eleven, equals nineteen. What is x? **6**

24 How many nought point twos are there, in two hundred? **1000**

25 Add nine squared to five squared. **106**

1 The capacity of a cup has been measured as nought point one nine litres, to the nearest one hundredth of a litre. Write down what the minimum capacity of the cup could be. 0.185 litres
2 Write thirty-three hundredths, as a percentage. 33%
3 What is plus eight, divided by minus two? −4
4 Write one quarter, as a decimal. 0.25
5 Write seven per cent, as a fraction. $\frac{7}{100}$
6 What is one seventh, divided by one fifth? $\frac{5}{7}$
7 Divide ninety-seven point nine by one hundred. 0.979
8 What is one quarter of one tenth? $\frac{1}{40}$
9 Write seven hundred thousandths, as a decimal. 0.7
10 What is nought point three, divided by nought point nought one? 30

11 Three angles of a quadrilateral are eighty degrees, seventy degrees, and eighty degrees. What is the size of the fourth angle? 130°
12 Write, as a percentage, seven, divided by fifty. 14%
13 Look at the equation on your answer sheet: y equals five x squared. Find y, when x is five. 125
14 What is ninety, divided by one thousand, eight hundred? 0.05
15 Increase a distance of twenty-metres by three fifths. 32 m
16 How far could you go in five hours, at a constant speed of seventy kilometres per hour? 350 km
17 What is nought point nought three, multiplied by nine hundred? 27
18 Write down, as a decimal to two decimal places, an estimate for the value of three hundred and four, divided by four hundred and four. (0.75)
19 Look at the expression on your answer sheet. What is the smallest whole number x could be? 19
20 How many fifty thousands are there, in one million? 20

21 Look at the expression on your answer sheet. Work out an approximate answer to this calculation. (200)
22 Find the mean of twelve, four, eight, three, seven, and two. 6
23 Look at the equation on your answer sheet: y equals thirty-six, minus three x. What is y, when x is nine? 9
24 Add two squared to nine squared. 85
25 A pond is being filled at the rate of eighteen litres per minute. Write this rate in litres per second. 0.3 litres/sec

1 What is the result of multiplying nought point two nought four by one thousand? 204
2 Write one fifth, as a decimal. 0.2
3 What is minus four, divided by minus two? 2
4 Write twenty-nine hundredths, as a percentage. 29%
5 Find the area of a square of side of length nine centimetres. 81 cm²
6 What is one third of one tenth? $\frac{1}{30}$
7 The capacity of a glass has been measured as six hundred and forty-two millilitres, to the nearest millilitre. What is the minimum capacity the glass could have? 641.5 ml
8 What is nought point nought nine, multiplied by nought point nought nought one? 0.00009
9 Write forty ten-thousandths, as a decimal. 0.004
10 What is nought point nought five, divided by nought point nought five? 1

11 At a garage, four cars out of ten, on average, fail the MOT test. Of seventy cars tested, how many might you expect to fail? 28
12 Look at the equation on your answer sheet: a equals c squared, plus five. Find a, when c is four. 21
13 Multiply nine thousand, two hundred by fifty. 460000
14 Increase an amount of three thousand dollars by thirty-five per cent. $4050
15 The perimeter of a rectangle is eighteen centimetres. The width of the rectangle is three centimetres. What is its area? 18 cm²
16 Write down, as a decimal to one decimal place, an estimate for the value of two hundred and twelve, divided by three hundred and eighty-six. (0.5)
17 What is nought point nought nought four, multiplied by three thousand? 12
18 Look at the expression on your answer sheet. What is the greatest whole number x could be? 5
19 Multiply nine hundred and eighty by seven hundred. 686000
20 A man cycles three kilometres in twenty minutes. What is his average speed for the journey? 9 km/hr

21 Estimate the value of forty-two per cent of four hundred and one. (160 or 168)
22 Take two squared from four squared. 12
23 Look at the equation on your answer sheet: five x, minus thirteen, equals seven. What is x? 4
24 How many nought point nought nines are there, in ninety? 1000
25 Look at the expression on your answer sheet. Work out an approximate answer to this calculation. (10)

Part 2: Answer Grids

In this section, photocopiable answer grids are provided for each test. The grids allow space for students to write their answers. For some questions, additional information is provided:

- to clarify numbers quoted in the questions

- to supply information referred to in the question, such as an illustration of an angle requiring an estimation of its size

- to reduce the need for students to retain numbers in their head at the same time as completing a mental calculation.

Students should have only a pen or a pencil; no other equipment, including anything that could be used to erase answers, should be available. Students should not be provided with any other means of calculating their working, such as calculators or additional paper.

Any working shown on a student's answer grid should invalidate the answer. However, students are permitted to cross out an answer. Space is then provided at the end of each grid for the student to write a question number and the new corresponding answer to that question.

The following guidance is given for the marking of answers.

Time

Answers related to questions about time are asterisked to make them distinguishable from a decimal answer.

A time could be given in terms of the 12- or 24-hour clock, or be written in words ('quarter to one'), any of which are acceptable. However, students should be encouraged to give answers in numbers rather than words since this will save them time.

In section E, the answer grids have space for time to be given in terms of hours and minutes, but if a student chooses to give the time in terms of fractions of an hour, then this should also be accepted.

Estimated measurements

Some questions require a student to indicate an estimated measurement on a diagram printed on the answer grid. Alternatively, some answers may require examination rather than a numerical check, for example, division of a line by bisectors. In these cases, the answer is given in brackets thus: (30°), (bisector).

It is suggested the following tolerances be applied to marking work: angles ±10°, lengths ±3mm. In section A, students are asked to divide rectangles and lines into halves or quarters. The lines of bisection should be given to within a tolerance of ±2mm.

Recurring decimals

In section E, students are asked to state the minimum or maximum a measure could be. The answer given beside the question is the limiting value. Answers which clearly state a recurring decimal are correct and should be accepted: for an answer given as 3.15, accept $3.14\dot{9}$.

Money amounts

If an answer is required in pence, p is shown as a guide on the answer grid, but so long as the answer includes the pound sign (£), an answer given in pounds is acceptable.

A1 Name: .... Date: ....

| 1 | | 11 | min | 21 | 39, 38, 56, 75, 47 |

| 2 | | 12 | | 22 | 32, 23, 31, 19 |

| 3 | | 13 | |

| 4 | | 14 | £ | 23 | 48, 58 | |

| 5 | | 15 | | 24 | 25p | p |

| 6 | | 16 | 60, 24 | | 25 | 400, 1530 | |

| 7 | | 17 | p |

| 8 | | 18 | 598 | |

| 9 | | 19 | |

| 10 | | 20 | 320, 64 | | **TOTAL MARKS** | |

- ✂ - - -

A2 Name: .... Date: ....

| 1 | | 11 | | 21 | 813, 633, 572, 754 |

| 2 | | 12 | |

| 3 | p | 13 | |

| 4 | | 14 | min | 22 | 47, 69, 36, 73, 24 |

| 5 | | 15 | | 23 | £3710, £500 | £ |

| 6 | | 16 | 25, 27 | | 24 | 37, 77 | |

| 7 | | 17 | £299, £10 | £ | 25 | 237 | |

| 8 | | 18 | |

| 9 | | 19 | £1.75 | p |

| 10 | | 20 | 44, 23 | | **TOTAL MARKS** | |

Name: .... **Date:** ....

1		11		21	48, 79, 57, 41, 34
2		12	m	22	269, 267, 176, 354
3		13			
4		14		23	87, 34
5		15	p	24	500, 1460
6	£	16	49, 24	25	54, 28
7		17			
8	hrs	18	52, 38		
9		19	398 km, 30 km / km		
10		20	30, 300		

TOTAL MARKS

---✂--------

Name: .... **Date:** ....

1	cm	11		20	398
2		12	min	21	730, 699, 823, 864
3		13	hrs		
4		14	£	22	20, 22, 13, 15, 25
5		15	———————	23	37, 76
6				24	1550
7		16		25	£3.50 / wks
8	p	17	29, 32		
9	£	18	57°, 14° / °		
10		19	£9.35 / p		

TOTAL MARKS

© Graham Newman 1997. Published by Thomas Nelson and Sons Ltd. Copyright permitted for purchasing schools only.

A5 Name: .... Date: ....

1	

2	

3	

4	yrs

5	

6	

7	hrs

8	days

9	

10	

11	

12	p

13	

14	

15	

16	25, 17	

17	£4.40	£

18	12, 46	

19	175 g	g

20	

21	22, 17, 13, 24, 26

22	45, 78, 54, 106

23	2830 mm 700 mm	mm

24	48, 54	

25	£2.99	£

TOTAL MARKS

- ✂ - - -

A6 Name: .... Date: ....

1	

2	

3	

4	

5	

6	hrs

7	

8	

9	

10	

11	min

12	

13	

14	°

15	hrs

16	37, 14	

17	sec

18	45, 55	

19	125 ml	ml

20	598	

21	45, 50, 46, 43, 47

22	258, 254, 245, 265

23	75, 76	

24	47 m, 29 m	m

25	234	

TOTAL MARKS

Name: .... **Date:** ....

1	
2	
3	
4	
5	
6	
7	
8	£
9	
10	

11	
12	
13	km
14	litres
15	min
16	p
17	£299, £69	£
18	30, 12	
19	47, 31	
20	

21	861, 531, 854, 165	
22	54, 46, 43, 53, 40	
23	89, 29	
24	41	
25	£810, £1437	£

TOTAL MARKS

-- ✂ --------

Name: .... **Date:** ....

1	
2	
3	
4	
5	£
6	
7	g
8	
9	
10	

11	kg
12	
13	
14	m
15	
16	
17	
18	33, 57	
19	£3.50	£
20	17, 32	

21	86, 68, 78, 64	
22	10, 30, 41, 13, 52	
23	254 km	km
24	72, 67	
25	£3800, £500	£

TOTAL MARKS

A9 Name: .... Date: ....

1	yrs
2	
3	
4	
5	
6	km
7	
8	
9	
10	

11	min
12	£
13	
14	litres
15	
16	45
17	£250, £190 £
18	22, 49
19	p

20	
21	44, 37, 18, 52, 33
22	145, 165, 232, 172
23	53, 27
24	£3.25 wks
25	59, 44

TOTAL MARKS

--✂-----

A10 Name: .... Date: ....

1	
2	
3	
4	
5	
6	
7	
8	
9	
10	°

11	
12	min
13	kg
14	
15	
16	72, 56
17	23, 39
18	£11.45 p
19	82, 18
20	395, 10

21	61, 55, 22, 66, 49
22	147, 193, 89, 235
23	54
24	39, 69
25	£1.99 £

TOTAL MARKS

A11 Name: .... Date: ....

1	

2	

3	

4	

5	£

6	hrs

7	

8	

9	

10	£

11	hrs

12	

13	

14	

15	min

16	698	

17	

18	42, 28	

19	£44	£

20	14, 29	

22	802, 799, 789, 832

21	17, 28, 14, 57, 21

23	89, 42	

24	£5.75, £3.50	£

25	132	

TOTAL MARKS [ ]

-- ✂ ------------

A12 Name: .... Date: ....

1	

2	

3	hrs

4	

5	

6	

7	

8	hrs

9	

10	

11	m

12	min

13	19, 5	cm

14	

15	£

16	40, 26	

17	38 kg, 14 kg	kg

18	199	

19	sec

20	33, 67	

21	46, 38, 51, 93, 47

22	91, 46, 127, 87

23	53, 37	

24	53	

25	67p, 38p	p

TOTAL MARKS [ ]

A13 Name: .... Date: ....

1	

2	

3	

4	

5	yrs

6	

7	hrs

8	

9	

10	

11	

12	days

13	min

14	

15	―――――――

16	47, 26	

17	p

18	33, 16	

19	£10, £5.50	£

20	18, 19, 20	

21	655, 541, 223, 684	

22	69, 24, 66, 31, 47

23	£2.80	wks

24	73, 46	

25	300, 1750	

TOTAL MARKS [ ]

--- ✂ ---

A14 Name: .... Date: ....

1	

2	

3	

4	

5	

6	

7	

8	

9	mm

10	

11	

12	

13	

14	

15	

16	60, 34	

17	225, 103	

18	693	

19	15p	p

20	21, 36	

21	465, 499, 456, 495	

22	54, 63, 10, 32, 51

23	83, 47	

24	£3.99	£

25	36	£

TOTAL MARKS [ ]

A15 Name: .... Date: ....

1	

2	

3	

4	

5	

6	mg

7	

8	

9	

10	

11	

12	_____

13	min

14	

15	

16	175 g	g

17	p

18	73, 27	

19	150 g	g

20	38, 54	

21	64, 58, 41, 30, 57

22	411, 421, 523, 510

23	83, 36	

24	235	

25	42, 68	

TOTAL MARKS

A16 Name: .... Date: ....

1	

2	

3	

4	

5	

6	

7	

8	

9	

10	

11	

12	

13	min

14	

15	

16	28, 47	

17	

18	£9.99	£

19	72, 53	

20	

21	55, 24, 71, 50, 48

22	261, 736, 465, 564

23	45, 87	

24	£2.35	wks

25	25	

TOTAL MARKS

A17 Name: .... Date: ....

1	

2	yrs

3	

4	

5	£

6	

7	

8	

9	

10	

11	

12	

13	1977	yrs

14	

15	days

16	122, 156	

17	£45, £26	£

18	34, 18	

19	95 cm	cm

20	

21	356, 84, 661, 254

22	68, 37, 31, 46, 67

23	1840	

24	78, 55	

25	25 g	g

TOTAL MARKS

---8<----

A18 Name: .... Date: ....

1	

2	

3	

4	

5	

6	hrs

7	£

8	

9	g

10	

11	

12	min

13	litres

14	

15	

16	34, 60	

17	£6.35	£

18	

19	

20	p

21	20, 32, 47, 14, 61

22	400, 398, 401, 389

23	83, 26	

24	£2.99	£

25	54, 46	

TOTAL MARKS

A19 Name: .... Date: ....

1	
11	
21	231, 123, 243, 112

| 2 | |
| 12 | |

3	
13	
22	63, 58, 49, 40, 32

4	
14	min
23	34, 76	

5	km
15	
24	2850 mm 700 mm	mm

6	
16	
25	44	

| 7 | |
| 17 | cm |

| 8 | m |
| 18 | 34, 58 | |

| 9 | |
| 19 | 29 p | £ |

| 10 | |
| 20 | 498 | |

TOTAL MARKS

A20 Name: .... Date: ....

1	
11	
21	58, 24, 31, 47, 20

2	
12	£
22	870, 895, 864, 887

| 3 | |
| 13 | litres |

4	
14	m
23	48, 37	

5	
15	min
24	£327	£

6	
16	597	
25	48	

| 7 | |
| 17 | wks |

| 8 | |
| 18 | 66, 33 | |

| 9 | |
| 19 | m |

| 10 | £ |
| 20 | |

TOTAL MARKS

A21 Name: .... Date: ....

1		**11**		**20**	27, 63		

2	

12	

21	55, 18, 53, 32, 11

3	

13	

22	360, 465, 326, 386

4	

14	

5	yrs

15	53°, 29°	°

23	2010

6	

16	

24	38, 76

7	cm

25	£4.10 wks

17	20, 483	

8	

18	

9	

19	£1.99, £1.49	p

10	

TOTAL MARKS

- ✂ - - - - - - -

A22 Name: .... Date: ....

1	p

11	

21	87, 61, 49, 38, 62

2	

12	litres

22	561, 210, 820, 653

3	

13	

4	p

14	min

23	44, 27	

5	

15	46, 25	

24	32 £

6	

16	

25	87, 29	

7	

17	12, 28	

8	

18	

9	

19	£4, £5	£

10	

20	

TOTAL MARKS

A23 Name: .... Date: ....

1	
2	
3	
4	yrs
5	
6	
7	
8	hrs
9	
10	km

11	
12	
13	
14	
15	min
16	
17	250 g	g
18	90, 46	
19	63 cm	cm
20	

21	34, 79, 63, 31
22	54, 65, 48, 51, 40
23	73, 27	
24	3650 ml	ml
25	316	

TOTAL MARKS

A24 Name: .... Date: ....

1	
2	
3	
4	
5	
6	
7	
8	
9	
10	

11	
12	
13	£
14	days
15	
16	p
17	
18	70 cm	cm
19	33, 54	
20	64, 37	

21	642, 851, 685, 645
22	61, 80, 41, 55, 64
23	74, 36	
24	37, 52	
25	43	

TOTAL MARKS

1		**11**		**20**	37, 29		
2		**12**		**21**	63, 61, 76, 53, 40		
3		**13**		**22**	265, 358, 199, 264		
4	£	**14**	min				
5	yrs	**15**	————————	**23**	64, 36		
6				**24**	£1.99	£	
7		**16**	59	**25**	56		
8		**17**					
9	£	**18**	485				
10		**19**	£4.35	p	**TOTAL MARKS**		

- ✂ - - - -

1		**11**	kg	**21**	687, 688, 668, 678		
2		**12**					
3		**13**		**22**	86, 61, 54, 42, 53		
4	litres	**14**	1992	yrs	**23**	56	
5		**15**	494	**24**	7480 mm 600 mm	mm	
6		**16**		**25**	59, 47		
7		**17**					
8	cm	**18**	sec				
9	p	**19**	18, 30				
10		**20**	68, 32	**TOTAL MARKS**			

1	**11** £	**20** 496
2	**12**	**21** 654, 633, 221, 335
3	**13** min	
4 £	**14**	**22** 27, 44, 72, 51, 45
5 p	**15** ――――――	**23** 88, 21
6		**24** £5.40 wks
7	**16** 19, 37	**25** 28, 45
8	**17**	
9 p	**18** 22, 68	
10	**19** 28, 41	**TOTAL MARKS**

--✂---------

1	**11**	**21** 40, 35, 74, 68, 27
2	**12** min	**22** 381, 172, 285, 297
3	**13**	
4	**14**	**23** 231
5	**15**	**24** 63 km, 49 km km
6	**16** 68, 38	**25** 38, 68
7	**17** 35 kg kg	
8 kg	**18** 35, 38	
9 hrs	**19**	
10	**20**	**TOTAL MARKS**

1	cm
2	
3	
4	
5	yrs
6	
7	
8	£
9	
10	

11	g
12	
13	
14	
15	
16	16, 38
17	38, 42
18	275 ml, 350 ml ml
19	497
20	p

21	58, 55, 50, 36, 27
22	687, 551, 364, 866
23	£6660, £320 £
24	44
25	64, 58

TOTAL MARKS

--✂----

1	
2	
3	
4	
5	p
6	
7	
8	£
9	
10	

11	
12	
13	°
14	min
15	
16	29, 34
17	493
18	g
19	17, 50

20	
21	45, 44, 52, 78, 21
22	846, 223, 561, 986
23	83, 28
24	252
25	84, 37

TOTAL MARKS

Name: .... **Date:** ....

1	
2	
3	
4	
5	
6	
7	
8	
9	
10	

11	
12	min
13	litres
14	kg
15	
16	p
17	33, 46
18	
19	21, 53
20	g

21	93, 85, 143, 742
22	76, 69, 66, 43, 70
23	£2940 £
24	79, 32
25	£35 £

TOTAL MARKS

Name: .... **Date:** ....

1	
2	
3	
4	
5	
6	
7	
8	
9	
10	

11	
12	°
13	
14	
15	———————————
16	12, 44
17	
18	37, 24
19	300 g g

20	30 days, 17th days
21	38, 49, 88, 22, 75
22	898, 897, 543, 856
23	58 km km
24	374
25	£3.75 wks

TOTAL MARKS

A33 Name: .... Date: ....

1	
2	
3	
4	
5	yrs
6	£
7	
8	£
9	
10	

11	
12	cm
13	
14	p
15	min
16	32, 21	
17	125 g	g
18	18, 30	
19	20	
20	£6.50	£

21	302, 813, 53, 934
22	19, 74, 71, 58, 62
23	53, 36	
24	63	
25	75, 46	

TOTAL MARKS

- ✂ - - -

A34 Name: .... Date: ....

1	
2	
3	
4	£
5	
6	
7	yrs
8	kg
9	
10	

11	
12	cm
13	min
14	
15	
16	496	
17	p
18	41, 29	
19	£1.99	p
20	29, 28	

21	982, 745, 888, 687
22	92, 71, 39, 67, 24
23	37, 74	
24	£2.99	£
25	7770 g, 400 g	g

TOTAL MARKS

A35 Name: .... Date: ....

1	
2	
3	
4	
5	
6	
7	
8	
9	
10	km

11	litres
12	min
13	days
14	
15	
16	39, 25
17	54, 46
18	30, 12
19	£3.25	£

20	
21	19, 74, 72, 77, 63
22	779, 977, 577, 978
23	53, 68
24	47
25	£4550	£

TOTAL MARKS

--✂------------------

A36 Name: .... Date: ....

1	
2	
3	
4	yrs
5	
6	hrs
7	
8	
9	
10	

11	°
12	litres
13	min
14	
15	
16	43, 27
17	692
18	sec
19	40, 24
20	£

21	806, 832, 866, 789
22	85, 48, 51, 93, 44
23	53, 39
24	264
25	67, 44

TOTAL MARKS

A37   Name: .... Date: ....

1	

2	

3	

4	

5	

6	

7	wks

8	

9	

10	

11	

12	

13	

14	min

15	_____

16	

17	109, 91	

18	393	

19	

20	13, 35	

21	202, 86, 365, 109	

22	44, 63, 92, 41, 64	

23	26 days	litres

24	75, 55	

25	1515 litres 700 litres	litres

TOTAL MARKS

--✂--------

A38   Name: .... Date: ....

1	

2	

3	

4	

5	£

6	£

7	

8	

9	

10	

11	km

12	

13	cm

14	£

15	

16	23, 40	

17	£8.60	£

18	495	

19	

20	

21	18, 23, 15, 21, 16	

22	683, 967, 564, 977	

23	£48, £19	£

24	48, 73	

25	£2.90	wks

TOTAL MARKS

Name: .... **Date:** ....

1	

2	

3	

4	

5	p

6	

7	hrs

8	

9	

10	

11	

12	

13	

14	min

15	———————

16	

17	

18	31, 64	

19	175 g	g

20	27, 36	

21	19, 16, 24, 35, 18

22	987, 986, 488, 897

23	48	£

24	36, 95	

25	£8840	£

TOTAL MARKS

Name: .... **Date:** ....

1	

2	

3	

4	yrs

5	

6	

7	

8	

9	

10	

11	

12	

13	

14	

15	min

16	24, 60	

17	£88	£

18	293	

19	

20	

21	590, 500, 531, 578

22	45, 24, 36, 41, 49

23	62, 29	

24	27	

25	87, 25	

TOTAL MARKS

81 Name: .... **Date:** ....

1	£
2	
3	
4	g
5	
6	
7	
8	
9	
10	£

11	£38, £17	£
12		p
13	24, 32, 11	
14	18 p	p
15	£	
16		
17	4.2, 3.1	
18		
19	112, 24	
20	%	

21	p	
22	£5.70, £3.50	£
23	640	
24	35, 11	
25	8:35	hr min

TOTAL MARKS

✂

82 Name: .... **Date:** ....

1	
2	p
3	
4	
5	
6	
7	
8	£
9	wks
10	£

11	28 cm, 33 cm	cm
12		
13	21, 14, 11	
14	14, 29	
15		
16	44	
17		
18	9.20	
19		
20	4, 3.2	

21	p	
22	78, 53	
23	390, 13	
24	£4.50	£
25	33, 13	

TOTAL MARKS

B3 Name: .... Date: ....

1	£
2	
3	
4	
5	litres
6	
7	
8	
9	
10	

11	235, 104	
12		
13	32, 14, 23	
14	893	
15		
16	min	
17	103, 36	
18		
19	1.1, 1.01	
20		

21	58 pound coins	pound coins
22	54, 46	
23	3.02, 4	
24		
25		

TOTAL MARKS

B4 Name: .... Date: ....

1	
2	
3	
4	
5	
6	
7	
8	
9	
10	

11	£34, £18	£
12	55, 26	
13	cm	
14	23, 14, 21	
15	156, 128	
16	41, 8	
17	%	
18	115, 26	
19	2.3, 0.14	
20		

21	p	
22	76, 34	
23	17, 510	
24	72, 11	
25	£20, £4.50	£

TOTAL MARKS

85 **Name:** .... **Date:** ....

1	hrs

2	

3	

4	

5	£

6	

7	

8	

9	

10	

11	p

12	58, 72	

13	18, 21, 30	

14	£9.99, £5.49	£

15	28, 47	

16	

17	1.1, 3.2	

18	

19	

20	5, 36	

21	6390 mm / 800 mm	mm

22	37, 58	

23	12, 240	

24	23, 15	

25	2 hrs 25 min	

TOTAL MARKS

---

86 **Name:** .... **Date:** ....

§	mm

2	

3	

4	

5	

6	pints

7	

8	

9	

10	

11	£

12	23, 20, 35	

13	498	

14	64, 25	

15	£

16	

17	1.3, 5.4	

18	

19	min

20	53, 9	

21	p

22	£200, £2804	£

23	

24	15, 450	

25	31, 12	

TOTAL MARKS

Name: .... **Date:** ....

1	

2	

3	

4	

5	

6	

7	

8	

9	£

10	

11	p

12	21, 20, 19	

13	£6.35, £2.40	£

14	

15	

16	%

17	62, 3	

18	%

19	43, 108	

20	3.1, 0.2	

21	700 m, 5500 m	m

22	

23	14 m, 280 m	

24	13, 13	

25	2 hrs 55 min	

TOTAL MARKS

Name: .... **Date:** ....

1	

2	

3	

4	hrs

5	

6	

7	m

8	days

9	

10	

11	m

12	64, 23	

13	25, 11, 21	

14	897	

15	

16	

17	2.3, 4.8	

18	22, 9	

19	%

20	13:20, 50 min	

21	£500, £1850	£

22	13, 520	

23	29, 87	

24	32, 11	

25	£6.25	£

TOTAL MARKS

B9 Name: .... Date: ....

1	
2	
3	
4	£
5	
6	
7	
8	
9	
10	

11	25, 28, 26	
12	32, 50	
13	583	
14	27, 73	
15	£1.99, £1.49	p
16	3.42, 1.21	
17		cm
18	34, 6	
18	48, 116	
20		

21		p
22	900, 1320	
23	2, 1.05	
24	12, 15	
25	18, 360	

TOTAL MARKS

- ✂ - - - -

B10 Name: .... Date: ....

1	£
2	
3	
4	
5	
6	
7	
8	
9	kg
10	

11	56 kg, 25 kg	kg
12		
13	791	
14		
15	£	
16		
17		
18	42	
19		
20		min

21	1750	
22		
23	1.75	
24	15, 750	
25	22, 13	

TOTAL MARKS

1	
2	
3	
4	
5	
6	
7	
8	
9	
10	

11	miles
12	
13	63 cm, 14 cm	cm
14	27, 21, 20	
15	
16	
17	2.9, 1.01	
18	
19	%
20	27, 106	

21	500, 1460	
22	21, 12	
23	71, 36	
24	13 : 40, 16 : 20	hrs min
25	16, 320	

TOTAL MARKS

1	
2	
3	
4	
5	
6	
7	
8	£
9	
10	

11	p
12	
13	19, 34	
14	596	
15	sec
16	32, 7	
17	3.4, 5.4	
18	£114, £51	£
19	
20	

21	300, 2010	
22	
23	2, 1.75	
24	52, 11	
25	£20, £4.50	£

TOTAL MARKS

B13 Name: .... Date: ....

1		
2		
3		
4	hrs	
5		
6		
7		
8		
9		
10		

11		
12	25, 20, 24	
13	64, 37	
14	cm	
15	33, 64	
16	67	
17	3.1	
18		
19	20 : 35, 21 : 15	mln
20	39, 118	

21	p	
22	360, 12	
23	47, 59	
24	£50, £12.50	£
25	21, 14	

TOTAL MARKS

B14 Name: .... Date: ....

1		
2		
3		
4		
5		
6		
7		
8		
9		
10		

11		
12	28, 51	
13		
14	68, 22	
15	996	
16		
17	4.8, 2.7	
18		
19	45 wks	days
20	%	

21	p	
22	400, 1330	
23	6, 2.75	
24	22, 13	
25	17, 340	

TOTAL MARKS

1	
2	
3	
4	£
5	
6	
7	
8	
9	
10	

11	
12	685	
13	
14	
15	
16	5.01, 3.04	
17	£4.35	£
18	4 cm	cm
19	55, 104	
20	20 : 30	

21	p
22	
23	
24	15, 600	
25	12, 11	

TOTAL MARKS

---✂-----------------

1	
2	
3	
4	
5	
6	
7	
8	
9	
10	

11	25 g	g
12	p
13	21, 63	
14	26, 33	
15	
16	2.5	
17	33, 8	
18	
19	1.2, 1.25	
20	min

21	hrs min
22	
23	11, 330	
24	31, 12	
25	£5.50, £30	£

TOTAL MARKS

817 Name: .... Date: ....

1	
2	
3	
4	kg
5	
6	hrs
7	
8	
9	
10	

11		
12	694	
13	20, 23, 25	
14	68, 32	
15	26	
16	sec	
17	3.4, 4.3	
18		
19	%	
20	47, 110	

21	p	
22	6.25	
23	64, 28	
24	22, 13	
25	12, 600	

TOTAL MARKS

---

818 Name: .... Date: ....

1	
2	
3	
4	
5	
6	
7	days
8	p
9	
10	

11		
12		
13	58, 35	
14	20, 23, 26	
15	kg	
16	4.5, 1.3	
17	57, 6	
18		
19	105, 38	
20		

21	£10.40, £40	£
22		
23	28, 54	
24	680, 40	
25	15, 15	

TOTAL MARKS

819 Name: .... Date: ....

1		**11**	g	**21** £3810, £500	£
2		**12**		**22** 4.25	
3		**13** 393		**23** 79, 32	
4		**14**		**24** 14, 700	
5		**15** 19, 33		**25** 21, 14	
6		**16** 43, 9		
7		**17** 0.31, 0.05		
8		**18** £112, £32	£	
9		**19**		
10		**20** min		**TOTAL MARKS**

--✂------------------------------

820 Name: .... Date: ....

1		**11** 38, 52		**21**	
2		**12** 48, 16		**22** 9, 7.25	
3		**13** p		**23** 19, 570	
4 °		**14** 797		**24** 47, 29	
5		**15** 275 ml 350 ml	ml	**25** 42, 12	
6		**16**		
7		**17**		
8 days		**18**		
9 min		**19** 55, 5		
10		**20** %		**TOTAL MARKS**

© Graham Newman 1997. Published by Thomas Nelson and Sons Ltd. Copyright permitted for purchasing schools only.

133

B21 Name: .... Date: ....

1		**11**		**21**	p	
2		**12**	23, 24, 25		**22**	
3		**13**	sec	**23**	11, 220	
4		**14**		**24**	21, 15	
5		**15**	£6.50	£	**25**	500, 1640
6		**16**				
7		**17**	5.42, 3.4			
8		**18**				
9		**19**				
10		**20**			**TOTAL MARKS**	

--- ✂ ---

B22 Name: .... Date: ....

1		**11**	g	**21**	£5.50, £30	£
2		**12**		**22**		
3		**13**	37, 24		**23**	22, 13
4	km	**14**		**24**	53, 27	
5	hrs	**15**		**25**	14, 420	
6		**16**	64, 3			
7		**17**	2.03, 1.3			
8		**18**	hrs			
9	p	**19**	45, 109			
10		**20**	min		**TOTAL MARKS**	

B23   Name: ....   Date: ....

#		#		#		
1		11	p	21	£86, £34	£
2	days	12		22	11, 440	
3		13	p	23		
4	yrs	14	29, 28		24	1.75	
5		15	696		25	21, 14	
6		16	5.42, 3.02			
7		17	%			
8		18	min			
9		19	35			
10		20	4.4, 4.04		TOTAL MARKS	

B24   Name: ....   Date: ....

#		#		#		
1	£	11	109, 91	21	°
2		12		22	12, 720	
3		13	21, 22, 23		23	hrs min
4		14	493		24	p
5		15		25	22, 15	
6		16				
7		17	5 cm / cm			
8		18	%			
9		19	21			
10		20	120, 35		TOTAL MARKS	

B25 Name: .... Date: ....

1		g
2	
3	
4		min
5	
6	
7		p
8	
9	
10	

11	54 g, 46 g	g
12		
13	25, 49	
14	£3.25	£
15		
16		
17		%
18	4.4, 2.3	
19		
20	38	

21	
22	
23	36, 83	
24	11, 770	
25	34, 12	

TOTAL MARKS

B26 Name: .... Date: ....

1		miles
2	
3	
4	
5	
6	
7	
8	
9	
10	

11	£43, £27	£
12	892	
13		
14		sec
15	£	
16	1.12, 3.04	
17	51	
18		
19	122, 85	
20	55 min	

21	2830 mm 700 mm	mm
22	26, 83	
23	61, 47	
24	16, 480	
25	21, 14	

TOTAL MARKS

827 Name: .... Date: ....

1	km	**11** 175 g	g	**21** p
2		**12** 20, 25, 24		**22** 11, 550
3		**13** 31, 74		**23** 15, 12
4		**14** 37, 36		**24** hrs min
5		**15**		**25** 82, 56
6		**16** 46		
7		**17**		
8		**18** 4.42, 1.02		
9		**19** 57, 124		
10		**20** %		**TOTAL MARKS**

--✂------------------------------

828 Name: .... Date: ....

1 hrs		**11**		**21** 44, 27
2		**12** 18, 20, 22		**22**
3		**13**		**23** 30, 540
4		**14** 595		**24**
5 £		**15** £8.40	£	**25** 15, 11
6		**16** 2.13, 4.23		
7		**17**		
8		**18** 101, 53		
9		**19** cm		
10		**20** 65		**TOTAL MARKS**

Name: .... **Date:** ....

1	
2	
3	
4	p
5	
6	
7	
8	
9	
10	

11	28 kg, 14 kg	kg
12		p
13		
14	774	
15	550, 135	
16		
17		
18	119, 53	
19	1.4, 2.51	
20		min

21	3640 ml 600 ml	ml
22	52, 37	
23	62, 45	
24	15, 900	
25	43, 11	

TOTAL MARKS

Name: .... **Date:** ....

1	
2	
3	
4	mm
5	
6	hrs
7	
8	
9	
10	

11	£	
12	18, 20, 22	
13		
14		
15	293	
16		
17		
18	%	
19	33, 102	
20		

21	64, 36	
22		
23	11, 880	
24	23, 13	
25		p

TOTAL MARKS

831 Name: .... Date: ....

1	

2	

3	

4	

5	

6	

7	

8	

9	

10	

11	£

12	698	

13	45, 55	

14	125 ml	ml

15	sec

16	

17	5.04, 1.02	

18	

19	31	

20	%

22	hrs min

22	83, 34	

23	6.25	

24	34, 12	

25	13, 260	

TOTAL MARKS [ ]

----------------------------✄---------

832 Name: .... Date: ....

1	

2	

3	

4	m

5	

6	

7	

8	

9	

10	

11	

12	p

13	

14	

15	£299, £79	£

16	37	

17	5.2, 3.02	

18	cm

19	42, 129	

20	%

21	37 cm, 76 cm	cm

22	

23	28, 45	

24	12, 480	

25	$3\frac{3}{4}$ hrs, 4 : 30	

TOTAL MARKS [ ]

Name: .... **Date:** ....

1	pints

11	43, 57	g

21	7450 mm 600 mm	mm

2	

12	

22	

3	

13	19, 21, 23	

23	11, 660	

4	

14	

24	25, 12	

5	

15	£3.50	£

25	48, 54	

6	

16	

7	

17	0.14, 0.25	

8	hrs

18	40 min	

9	

19	121, 37	

10	

20	

TOTAL MARKS

Name: .... **Date:** ....

1	hrs

11	

21	

2	

12	

22	35, 12	

3	

13	

23	£25, £6.80	£

4	

14	495	

24	63, 49	

5	

15	£11.45	p

25	19, 380	

6	

16	

7	

17	

8	

18	min

9	

19	54	

10	

20	

TOTAL MARKS

B35 Name: .... Date: ....

#			#			#		
1			11	£250, £199	£	21		
2			12	18, 20, 22		22	13, 650	
3			13			23	84, 37	
4	litres		14	69, 22		24	11 : 30, 4 : 15	hrs min
5			15	p		25	56, 11	
6			16					
7			17	6.9 2.4				
8	wks		18					
9			19	63				
10	£		20	52, 107		**TOTAL MARKS**		

- ✂ - - - - - - - -

B36 Name: .... Date: ....

#			#			#		
1			11			21	p	
2			12			22	72, 67	
3			13	18, 76		23	£2940, £200	£
4			14	£2.80	£	24	64, 11	
5			15			25	£6.50, £30	£
6			16					
7			17	61				
8			18	2.71, 3.04				
9			19	min				
10	yrs		20	117, 58		**TOTAL MARKS**		

837 Name: .... Date: ....

1		
2		
3		
4		
5	cm	
6		
7		
8		
9		
10		

11	£	
12	898	
13		
14	19, 14	
15		
16		
17		
18	%	
19	52	
20	45 min	

21	£9.40, £30	£
22	9, 5.75	
23	36, 53	
24	44, 59	
25	22, 14	

TOTAL MARKS

---

838 Name: .... Date: ....

1		
2		
3		
4		
5		
6		
7		
8		
9		
10		

11	£	
12	p	
13	57, 26	
14		
15	18, 19, 20	
16	5.12, 2.21	
17		
18	52 wks	wks
19	113, 34	
20	min	

21	7770 g, 400 g	g
22		
23	11, 330	
24	17, 12	
25	£6.50 £30	£

TOTAL MARKS

B39 Name: .... Date: ....

1	hrs	**11**		**21** £4515, £600	£
2		**12** 26, 60		**22** 89, 42	
3		**13** sec		**23** 16, 800	
4	km	**14** 33, 67		**24** 58, 11	
5		**15** 299		**25** £50, £11.50	£
6		**16** 24, 38			
7		**17** 44, 107			
8	days	**18**			
9		**19** 0.12, 0.21			
10		**20** 50 min		**TOTAL MARKS**	

- ✂ - - - - - - - - -

B40 Name: .... Date: ....

1		**11** 175 g	g	**21** 1515 litres 800 litres	litres
2		**12** 23, 21, 19		**22** 67, 38	
3		**13** p		**23** 17, 850	
4		**14** 73, 27		**24** $4\frac{1}{4}$ hrs, 3 : 30	
5		**15** £28, £64	£	**25** 17, 13	
6		**16**			
7		**17**			
8	wks	**18** 47			
9		**19**			
10		**20** %		**TOTAL MARKS**	

1	
2	
3	
4	
5	
6	
7	
8	
9	m
10	lb

11	%
12	
13	58 £
14	0.03, 0.12
15	kg 2 3 kg
16	km
17	£250 £
18	704 kg kg

19	212
20	°
21	
22	4.25
23	240
24	650 km 465 km km
25	3270

TOTAL MARKS

--- ✂ ---

1	
2	
3	
4	wks
5	cm
6	
7	
8	kg
9	
10	°

11	50 min
12	127, 59
13	4.2
14	29
15	cm
16	287, 36
17	cm _____
18	
19	123

20	hrs
21	£5.50, £20 £
22	456, 854
23	16, 960
24	1388
25	22, 11

TOTAL MARKS

C3 Name: .... Date: ....

1		**11**	24	**20**	25%, 80
2		**12**		**21**	£5.50, £20 £
3		**13**		**22**	15, 450
4		**14**	7.47, 0.13	**23**	4.5, 10.5
5		**15**	(angle diagram) °	**24**	4208
6	°			**25**	327, 896
7	cm	**16**	55, 303		
8	£	**17**	1¾ hrs min		
9		**18**	yds		
10	litres	**19**	321	**TOTAL MARKS**	

--✂--------------

C4 Name: .... Date: ....

1		**11**		**20**	512 £
2	p	**12**		**21**	3465
3		**13**	42	**22**	236, 886
4		**14**	9.14, 7.03	**23**	12, 840
5		**15**	%	**24**	2.3, 4.3
6		**16**	A ——————— B	**25**	(map with A, B, C) km
7	cm				
8		**17**	68, 893		
9		**18**	350g g		
10	km	**19**	hrs	**TOTAL MARKS**	

1	
2	
3	
4	
5	
6	
7	mm
8	
9	
10	kg

11	min
12	
13	2.2, 1.8	
14	m

15	53	
16	45, 224	
17	800	
18	$1\frac{1}{3}$ hrs	min

19	
20	

0°

21	884, 356	
22	3255	
23	
24	4.25	
25	11, 330	

TOTAL MARKS

1	m
2	wks
3	
4	
5	
6	
7	
8	
9	°
10	ml

11	
12	5.01, 2.2	
13	31, 126	
14	46	
15	
16	40	
17	709, 54	
18	£90	£
19	404	
20	210 min	hrs

21	6.5, 12.5	
22	13, 390	
23	hrs min	
24	872, 248	
25	5976	

TOTAL MARKS

C7 Name: .... Date: ....

1 £	
2	
3	
4	
5	
6	
7 cm	
8	
9 mm	
10 g	

11	28, 125	
12	5.8	
13		
14	44 yrs	yrs
15		°
16	120 gal	gal
17	483, 76	
18		
19	123	

20	$3\frac{1}{4}$ hs	min
21	2492	
22	46, 11	
23		km
24	12, 480	
25	294, 937	

TOTAL MARKS

--✂----------

C8 Name: .... Date: ....

1	
2	
3	
4	
5	
6 litres	
7	
8	
9 m	
10 in	

11		%
12	5.3, 2.4	
13		
14	amps	amps
15	67, 504	
16		
17	10%, 30	
18	72 ml	ml

19		mm
20	313	
21	6.5, 8.5	
22	600	
23	£25, £6.50	£
24	4068	
25	546, 675	

TOTAL MARKS

C9 Name: .... Date: ....

1	

2	

3	

4	

5	

6	cm

7	

8	°

9	

10	kg

11	cm

12	

13	0.33, 2.3	

14	£113, £46	£

15	36	

16	210 g	g

17	397, 49	

18	3600 lire	lire

19	$1\frac{1}{4}$ hrs	min

20	0°

21	£15.50	£

22	76, 11	

23	13, 260	

24	6032	

25	459, 752	

TOTAL MARKS [ ]

C10 Name: .... Date: ....

1	£

2	

3	

4	

5	

6	mm

7	£

8	°

9	litres

10	

11	min

12	1.1	

13	

14	

15	64	

16	75% 8000 lire	lire

17	55, 789	

18	214	

19	27	

20	0°

21	888, 417	litres

22	16, 480	

23	11, 15	

24	11 : 15, 15 : 20	hrs min

25	3184	

TOTAL MARKS [ ]

Name: .... **Date:** ....

1		hrs
2	
3	
4	
5	
6	
7		g
8	
9		cm
10		km

11	303, 37	
12		
13		
14	49, 104	
15		°C

16	5.24, 2.12	
17		g
18	20%, 20	

19	413	
20	270 min	hrs
21	788, 526	
22	14, 560	
23	4086, 9	
24	4.5, 12.5	
25		km

TOTAL MARKS

Name: .... **Date:** ....

1		wks
2	
3	
4	
5	
6		cm
7		°
8	£
9	
10		litres

11		
12		%
13	4.7, 2.13	
14	480	
15	16	
16		
17		days
18	89, 218	
19	$3\frac{3}{4}$ hrs	min

| 20 | |

21	10 : 50, $4\frac{1}{4}$ hrs	
22	12, 360	
22	34, 11	
24	686, 535	
25	2250	

TOTAL MARKS

C13 Name: .... Date: ....

1		11	%	20	135 min	hrs

1

11 %

20 135 min | hrs

2

12

21 £5.50, £20 | £

3

13 52

22 14, 280

4

14 109, 43

23 66, 11

5

15 °

24 4104, 9

6 cm

25 842, 498

7

16 950 g | g

8

17 96

9

18 312

10 km

19 507, 64

TOTAL MARKS

- ✂ - - - -

C14 Name: .... Date: ....

1

11 £433, £75 | £

19 2¾ hrs | min

2

12

20 $80 | $

3

13 3.04. 1.05

21 220, 11

4

14 kg | 20 30 kg

22 3.5, 13.5

5

23 943, 388

6

24 3872

7 m

15 14 | hrs

25 | kg

8

16 414

9

17 75%, 4000 pts | pts

10 g

18 103, 54

TOTAL MARKS

© Graham Newman 1997. Published by Thomas Nelson and Sons Ltd. Copyright permitted for purchasing schools only.

150

1		11		20	

2		12	

3		13	3.1, 3.1		21	672, 569	

4		14	62		22	15, 600	

5		15			23	16 : 40, 4 hrs 55 min	

6		16	550 kg	kg	24	48, 11	

7		17	294, 48		25	4606	

8	mm	18	£120	£

9		19	$1\frac{1}{3}$ hrs	min

10	litres

TOTAL MARKS [ ]

---- ✂ ----

1		11	cm	21	574, 647	

2		12		22	22, 13	

3		13	26		23	16, 320	

4		14	37, 106		24	6813	

5		15	min	25	km

6	ml	16	430		

7		17	800 gal	gal

8	ft	18	9000 pts	pts

9	min	19	

10	km	20	903, 28	cm

TOTAL MARKS [ ]

C17 Name: .... **Date:** ....

1	

2	

3	

4	

5	

6	cm

7	

8	

9	°

10	g

11	%

12	

13	

14	2.75, 0.05	

15		mm

16	817, 58	

17	213 wks	days

18	25%, 40	

19	200	

20	A ———————————————— B

21	3.8, 5.8	

22	13, 650	

23	£100, £25.50	£

24	3504, 6	

25	663, 567	

TOTAL MARKS [ ]

- ✂ - - -

C18 Name: .... **Date:** ....

1	

2	

3	

4	

5	

6	litres

7	

8	°

9	

10	cm

11	56 km, 102 km	km

12	17	

13	0.15, 0.15	

14		°

15	

16	57, 444	

17	212	

18	750 ml	ml

19	$1\frac{1}{4}$ hrs	min

20	60 mm	mm

21	767, 585	

22	14, 420	

23	£100, £26	£

24	3800	

25	44, 11	

TOTAL MARKS [ ]

C19 Name: .... Date: ....

1	
2	
3	
4	
5	
6	mm
7	
8	£
9	
10	km

1	min	
12	5.03, 2.4	
13		
14	57	
15		
16	400 kg	kg
17		
18	415	
19	283, 38	
20	150 min	hrs

21	435, 798	
22	5202	
23	6.5, 8.5	
24	340, 17	
25		km

TOTAL MARKS

C20 Name: .... Date: ....

1	
2	
3	
4	
5	
6	m
7	g
8	
9	
10	

11		
12		
13	3.2, 1.04	
14	65	
15	209	
16	$3\frac{2}{3}$ hrs	min
17	60g	g
18	43, 578	
19		cm

20	2.5	
21	3828	
22	15, 11	
23		hrs min
24	18, 540	
25	£887, £345	£

TOTAL MARKS

© Graham Newman 1997. Published by Thomas Nelson and Sons Ltd. Copyright permitted for purchasing schools only.

153

C21 Name: .... Date: ....

1			11	min		20	$2\frac{1}{3}$ hrs	min
2			12	1.3, 2.6		21	846, 477	
3			13	45		22	16, 320	
4	p		14			23	14 : 40, 17 : 20	hrs min
5			15	(angle diagram) °		24	4185	
6			16	250 ml ml		25	22, 12	
7			17	203, 48				
8	cm		18	123				
9	kg		19	120				
10	ml							

TOTAL MARKS ☐

--------------------------------✂------------------------------------

C22 Name: .... Date: ....

1	km		11	%		20	(angle diagram) 0°
2			12			21	£2910 £
3			13			22	53, 11
4			14	28, 105		23	12, 840
5			15			24	£20, £4.50 £
6	cm		16	124		25	878, 345
7			17	$2\frac{1}{3}$ hrs min			
8	°		18	10%, 50			
9	litres		19	£609, £57 £			
10							

TOTAL MARKS ☐

C23 Name: .... Date: ....

1		11	600 m ___ m	19	287, 76
2		12	47, 101	20	135 min ___ hrs
3		13		21	7.5, 15.5
4		14	___ %	22	360° ___ °
5		15	___ g	23	837, 584
6	___ m		5.7 / 5.8 grams	24	6055
7	£	16	2.1, 3.5	25	___ km
8		17	£90 ___ £		A — B — C
9	___ g	18	323		
10	___ °				**TOTAL MARKS**

C24 Name: .... Date: ....

1	£	11	___ %	20	250 g ___ g
2		12	2.16, 0.04	21	£50, £12.50 ___ £
3		13		22	4.2, 6.2
4		14	73	23	17, 340
5		15	£43, £109 ___ £	24	804, 357
6	___ °	16	___ mm	25	3912
7			A _____ B		
8	___ km	17	341		
9	___ km	18	6000 lire ___ lire		
10	___ lb	19	58, 408		**TOTAL MARKS**

© Graham Newman 1997. Published by Thomas Nelson and Sons Ltd. Copyright permitted for purchasing schools only.

155

C25 Name: .... Date: ....

1		

2		

3		wks

4		

5		

6		mm

7	£	

8		°

9		

10		litres

11		cm

12		

13		

14	48, 117	miles

15		km/hr

km/hr
21
20 22

16	40	

17	79, 303	

18	5%, 3000 kg	kg

19	404	

20	$2\frac{1}{4}$ hrs	min

21	747, 596	

22	12, 600	

23	£5.50, £30	£

24	22, 13	

25	6208	

TOTAL MARKS

- ✂ - - -

C26 Name: .... Date: ....

1		

2		

3		

4		

5		

6		kg

7		

8		

9		

10		cm

11		min

12	3.07, 1.3	

13		

14		°

15	25 p	£

16	47, 319	

17	50%, 70	

18	302	

19	330 min	hrs

20	70 kg	kg

21	3.5, 9.5	

22	4095	

23	£10.40, £40	£

24	680, 40	

25	£856, £567	£

TOTAL MARKS

C27 Name: .... Date: ....

1	

2	days

3	

4	

5	

6	km

7	°

8	

9	m

10	£

11	

12	8.12, 3.01

13	37

14	%

15	

16	59, 572

17	$3\frac{1}{3}$ hrs	min

18	55 miles	miles

19	

20	242

21	347, 598

22	700, 14

23	£5193	£

24	5.2, 7.2

25	km

C, A, B

TOTAL MARKS

C28 Name: .... Date: ....

1	

2	

3	

4	

5	p

6	£

7	cm

8	

9	g

10	

11	114, 39

12	3.01, 1.33

13	18 : 35, 19 : 25	min

14	

15	%

16	350

17	96 kg	kg

18	0°

19	305

20	£350, £74	£

21	3 hrs 10 min 11 : 20	

22	19, 570

23	43, 12

24	4068

25	562, 689

TOTAL MARKS

C29 Name: .... Date: ....

1	£

2	

3	

4	

5	

6	litres

7	

8	kg

9	m

10	

11	

12	

13	28	

14	

15	

16	£150	£

17	24 hrs	hrs

18	154	

19	A |————————————————| B

20	£389, £89	£

21	487, 637	

22	14, 420	

23	£5.50, £30	£

24	7.5, 15.5	

25	3952	

TOTAL MARKS

C30 Name: .... Date: ....

1	

2	

3	

4	

5	

6	cm

7	pints

8	

9	

10	litres

11	9.33, 3.03	

12	55, 104	

13	55 min 20 : 30	

14	£77	£

15	mm

4.2

mm   4.3

16	68 g	g

17	302	

18	

19	4000	

20	$1\frac{1}{3}$ hrs	min

21	hrs min

22	664, 548	

23	11, 440	

24	3.7, 7.7	

25	6714	

TOTAL MARKS

C31 **Name:** .... **Date:** ....

1	wks
2	
3	
4	
5	
6	°
7	cm
8	
9	kg
10	m

11	cm
12	7.51, 1.11
13	43
14	
15	85, 123
16	234
17	210 min	hrs
18	£75, £503	£
19	180
20	24

21	3.5, 9.5
22	720
23	547, 697
24	6139
25	km

TOTAL MARKS

--------------------------------✂------------------

C32 **Name:** .... **Date:** ....

1	
2	
3	
4	
5	
6	£
7	cm
8	
9	
10	km

11	min
12	5.78, 3.72
13	32
14	110, 48
15	°

16	400
17	312
18	25%, 100
19	587, 68

20	$2\frac{1}{3}$ hrs	min
21	4488
22	586, 654
23	$4\frac{1}{4}$ hrs, 10 : 05
24	11, 770
25	4.5, 14.5

TOTAL MARKS

C33 Name: .... Date: ....

1		11	%	20	A ├───────────────────┤ B
2		12		21	hrs min
3		13	72	22	16, 12
4		14		23	11, 550
5		15	120	24	5312
6	litres	16	55	25	775, 468
7		17	401		
8		18	203, 54		
9	°	19	550		
10					**TOTAL MARKS**

- ✂ - - - -

C34 Name: .... Date: ....

1		11	40 min	20	$3\frac{3}{4}$ hrs min
2		12	2.37, 5.02	21	4 hrs 20 min
3		13	74	22	540
4		14	£105, £38 £	23	12, 15
5		15	°	24	356, 868
6	cm	16	330	25	3512
7		17	314 g g		
8		18	609, 76		
9	km	19	39		
10	ft				**TOTAL MARKS**

C35 Name: .... Date: ....

1

2

3

4

5

6

7 kg

8 °

9 p

10 cm

11 67

12 34, 114

13

14 7.54, 2.44

15

16

17 411

18

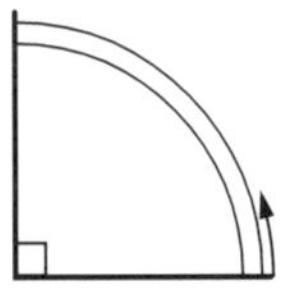

19 350 miles | miles

20 47, 692

21 4606 days | wks

22 44, 11

23 15, 900

24 655, 786

25 km

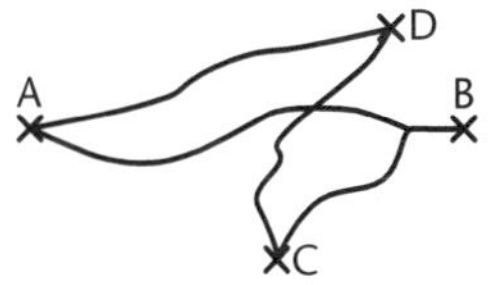

TOTAL MARKS

C36 Name: .... Date: ....

1 wks

2

3

4

5

6

7

8 m

9

10 litres

11 %

12

13 34

14 108 litres
45 litres | litres

15 litres

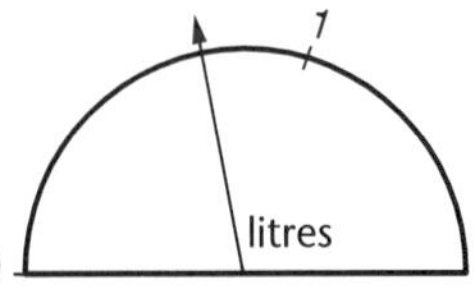

16 75%, £4000 | £

17 316

18 120

19 $1\frac{1}{3}$ hrs | min

20 £804, £65 | £

21 7.4, 9.4

22 13, 260

23 11 : 40, 16 : 50 | hrs min

24 354, 878

25 6012

TOTAL MARKS

Name: .... **Date:** ....

1		**11** 18 : 30, 19 : 10 min	**20** £450 £
2		**12**	**21** 507, 684
3		**13** 3.33, 6.05	**22** 19, 380
4		**14** 63	**23** 31, 12
5		**15**	**24** £25, £6.50 £
6 mm		**16** 87, 621	**25** 6280
7		**17** 316	
8 °		**18** mm	
9 km		A _____ B	
10 km		**19** 96 gal gal	**TOTAL MARKS**

Name: .... **Date:** ....

1 wks		**11** 36 cm, 120 cm cm	**19** 90
2		**12**	**20** $3\frac{2}{3}$ hrs min
3		**13** 3.51, 1.11	**21** 904, 357
4		**14** kg	**22** 13, 650
5			**23** 57, 11
6 kg		**15** min	**24** hrs min
7		**16** 503, 54	**25** 5268
8 °		**17** 10%, 700	
9		**18** 421, 7	
10 kg			**TOTAL MARKS**

C39 **Name:** .... **Date:** ....

1		**11** £140, £27	£	**20** 100 min	hrs
2		**12** 35		**21** 996, 239	
3		**13**	%	**22** 16, 800	
4		**14** 3.02, 2.22		**23** £7.50, £50	£
5		**15**	°	**24** 4809	
6	m	**15** (angle diagram)		**25** 7.4, 11.4	
7	min	**16** 403			
8		**17** 370 ft	ft		
9		**18** 63, 798			
10	ml	**19** 24		**TOTAL MARKS**	

- ✂ - - - - - -

C40 **Name:** .... **Date:** ....

1	yrs	**11**	cm	**20** 5%, £800	£
2		**12** 85, 123		**21** 6224	
3		**13** 3.07, 1.02		**22** 11, 330	
4		**14** 18		**23** £485, £847	£
5		**15**	min	**24** 3.7, 7.7	
6		**16** 427		**25**	km
7		**17**	m	(map with points A, B, C, D)	
8		**18** (angle diagram) 0°			
9	km				
10		**19** 303, 98		**TOTAL MARKS**	

D1 Name: .... Date: ....

1	cm
2	g
3	
4	
5	
6	
7	5.18
8	
9	
10	%

11	$1\frac{3}{4}$ hrs — min
12	30 g — g
13	25%, 20
14	203
15	0°
16	80°, 90°, 100° — °
17	$y = 13 - 2x$
18	510 g — g
19	$\frac{1}{8}$, 2.5

20	30
21	km
22	£480, 16 — £
23	
24	$2x - 3 = 4$
25	59, 11

TOTAL MARKS

- ✂ - - - -

D2 Name: .... Date: ....

1	0.7 — mm
2	cm
3	7.45
4	
5	
6	
7	cm²
8	13.3
9	%
10	

11	£491 — £
12	90 kg — kg
13	
14	$2\frac{2}{3}$ hrs — min
15	123
16	cm²
17	
18	$C = 2(d + 3)$
19	
20	£64 — £

21	247, 398
22	360
23	16, 12
24	21
20	

TOTAL MARKS

D3 Name: .... Date: ....

1		m

| 2 | 7.4 | ° |

| 3 | |

| 4 | | ml |

| 5 | |

| 6 | | % |

| 7 | 3.07 | |

| 8 | | % |

| 9 | |

| 10 | |

| 11 | 241 km | km |

| 12 | | cm |

| 13 | 250 | £ |

| 14 | 302 | |

| 15 | | cm |

A ― B

| 16 | 80°, 60° | ° |

| 17 | 9 : 2 | |

| 18 | 440 | |

| 19 | 220 litres | litres |

| 20 | $p = (q+2)^2$ | |

| 21 | 5.7, 15.7 | |

| 22 | 280 | |

| 23 | 876, 6 | |

| 24 | 3, 5, 9, 11 | |

| 25 | $y = 6x - 14$ | |

TOTAL MARKS [ ]

--✂------------

D4 Name: .... Date: ....

| 1 | | km |

| 2 | | cm |

| 3 | |

| 4 | |

| 5 | |

| 6 | |

| 7 | 9.81 | |

| 8 | | % |

| 9 | | m² |

| 10 | |

| 11 | 107 | |

| 12 | | in |

| 13 | 512 | |

| 14 | | £ |

| 15 | 48 litres | litres |

| 16 | 110°, 120°, 80° | ° |

| 17 | 6 | |

| 18 | 38 m | m |

| 19 | $a = 2c^2$ | |

| 20 | |

| 21 | | km |

| 22 | 13, 12 | |

| 23 | 30, 540 | |

| 24 | |

| 25 | $5x + 9 = 24$ | |

TOTAL MARKS [ ]

D5 Name: .... Date: ....

1	m
2	sec
3	
4	km
5	
6	
7	87.9
8	
9	%
10	m²

11	£56 £
12	394
13	60
14	
15	0°
16	
17	20%, 50
18	34 cm cm²
19	%

20	$y = x^2 - 4$
21	£857, £475 £
22	17, 15
23	900
24	
25	5, 1, 1, 3, 5

TOTAL MARKS

--✂----

D6 Name: .... Date: ....

1	kg
2	
3	
4	2 gallons litres
5	5.3 litres ml
6	
7	
8	%
9	19.1
10	

11	203 £
12	
13	°
14	120 mg mg
15	$1\frac{1}{4}$ hrs min
16	
17	3 : 2 g
18	70°, 65° °
19	

20	$a = \dfrac{24}{c + 3}$
21	8
22	220
23	$y = 4x - 16$
24	549, 9
25	14, 12

TOTAL MARKS

D7 Name: .... Date: ....

1	

2	

3	°

4	litres

5	cm

6	%

7	0.022	

8	m²

9	

10	0.85	

11	709	

12	126	

13	$2\frac{1}{3}$ hrs	min

14	50	g

15	

16	

17	92	£

18	$d = 4(e - 2)$	

19	16 cm	cm²

20	

21	600	

22	3.5, 11.5	

23	485, 746	

24	2, 4, 2, 7, 5	

25	

TOTAL MARKS [ ]

--✂--------------

D8 Name: .... Date: ....

1	yrs

2	m

3	°

4	

5	51.2 g	kg

6	

7	78.5	

8	%

9	

10	

11	800 m	m

12	187	

13	£90	£

14	232	£

15	0°

16	

17	80°, 110°, 90°	°

18	$y = 6x^2$	

19	520	

20	120 ml	ml

21	A ... B	km

22	28, 11	

23	260	

24	

25	$6x - 12 = 30$	

TOTAL MARKS [ ]

© Graham Newman 1997. Published by Thomas Nelson and Sons Ltd. Copyright permitted for purchasing schools only.

D9 Name: .... Date: ....

2	km
1	
3	°
4	
5	lb
6	m²
7	
8	4.99
9	%
10	

11	60°, 90° °
12	405
13	cm A _____ B
14	g
15	
16	litres
17	
18	
19	$d = (e + 3)^2$

20	3 : 7 cm
21	7.5, 9.5
22	380, 19
23	976
24	
25	$y = 8x - 18$

TOTAL MARKS

- ✂ - - -

D10 Name: .... Date: ....

1	days
2	mm
3	
4	°
5	litres
6	
7	0.392
8	
9	%
10	0.47

11	
12	803
13	4000
14	504
15	$2\frac{1}{4}$ hrs min
16	10 m m²
17	%
18	
19	$y = x^2 + 3$
20	36

21	12, 16
22	16, 320
23	km
24	
25	$7x + 15 = 50$

TOTAL MARKS

D11 Name: .... Date: ....

1	m²

11	219	sec

20		

2	

12	70 kg	kg

21	6.5, 10.5	

3	kg

13	203	

22	13 miles / 650 miles	gal

4	

14	50%, 60	

23	847, 378	

5	cm

15	A ⊢————————⊣ B

24	16	

6	%

16	£

25	

7	

17	c = 25 − 3x	

8	7.46	

18	20 g	g

9	

19	10 cm	cm²

10	0.45	%

TOTAL MARKS [ ]

D12 Name: .... Date: ....

1	km

11	100°, 130°, 70°	°

21	861 days	wks

2	

12	3⅓ hrs	min

22	25, 650	

3	°

13	492	

23	15, 15	

4	m

14	55 miles	miles

24	2, 5, 3, 8, 9, 3	

5	

15	142	

25	y = 7x − 14	

6	

16	2700	

7	51.6	

17		

8	m²

18	$d = \dfrac{16}{c + 3}$	

9	%

19	3 : 2	£

10	

20		

TOTAL MARKS [ ]

D13 Name: .... Date: ....

1 _____ days	**11** 96 _____ £	**20** 40 m _____ m
2 _____ cm	**12** 202 _____	**21** 800, 16 _____
3 _____	**13** _____ kg	**22** 46, 11 _____
4 84.5 kg _____ g	**14** 205 _____	**23** 886, 475 _____
5 _____	**15** 0°	**24** _____
6 _____		**25** $6x - 11 = 37$ _____
7 63.6 _____	**16** _____	
8 _____	**17** $p = (q + 3)^2$ _____	
6 _____ %	**18** _____	
10 _____ %	**19** 8 cm _____ cm²	**TOTAL MARKS** _____

D14 Name: .... Date: ....

1 _____ litres	**11** _____ hrs	**20** _____ g
2 _____	**12** £170 _____ £	**21** 687 _____
3 _____ m	**13** 245 _____	**22** 420, 14 _____
4 _____	**14** 489 _____	**23** 22, 11 _____
5 _____	**15** _____ mm A ├─────┤ B	**24** _____
6 _____	**16** 85°, 45° _____ °	**25** 8, 12, 8, 16 _____
7 4.41 _____	**17** _____	
8 _____ cm²	**18** _____ £	
9 _____ %	**19** $y = 3(x - 2)$ _____	
10 0.45 _____		**TOTAL MARKS** _____

D15 Name: .... Date: ....

1	cm
2	
3	litres
4	
5	pints
6	
7	13.01
8	%
9	m^2
10	

11	
12	$1\frac{1}{3}$ hrs	min
13	£68	£
14	
15	207
16	12 m^2	m
17	
18	$d = 3e^2$
19	
20	4 : 3

21	32
22	3.5, 5.5
23	km

| 24 | 340, 17 |
| 25 | $y = 8x - 16$ |

TOTAL MARKS

---✂-----------

D16 Name: .... Date: ....

1	°
2	
3	kg
4	
5	m
6	0.038
7	
8	
9	
10	%

11	24	km
12	324
13	180 mg	mg
14	603
15	

A ├───────────────────┤ B

16	24 mm	mm^2
17	
18	$a = \dfrac{12}{c + 2}$
19	£2	£

20	12
21	876, 698
22	330
23	12, 24
24	
25	$5x - 15 = 20$

TOTAL MARKS

D17 Name: .... Date: ....

1	km
2	
3	
4	cm
5	
6	%
7	83.1
8	m²
9	%
10	

11	£
12	mm
13	487
14	
15	0°
16	10%, 30
17	
18	$y = x^2 + 4$
19	80°, 90°, 80° °

20	
21	868
22	16, 800
23	76, 11
24	3, 9, 5, 3
25	

TOTAL MARKS

D18 Name: .... Date: ....

1	litres
2	
3	
4	°
5	£
6	
7	
8	%
9	3.37
10	

11	£203 £
12	ml
13	
14	
15	A ———————— B
16	
17	80 g g
18	24 cm²
19	$p = 20 - 4q$

20	4 : 3
21	5.5, 9.5
22	18, 540
23	846, 467
24	32
25	$y = 8x - 12$

TOTAL MARKS

D19 Name: .... Date: ....

1	cm

2	

3	km

4	

5	90 cm	ft

6	0.755	

7	

8	m²

9	%

10	

11	214	

12	330	

13	509	

14	$

15	$3\frac{1}{4}$ hrs	min

16	20	

17	100°, 80°, 100°	°

18	250	

19	77 kg	kg

20	$y = \dfrac{20}{x+2}$	

21	4.6, 6.6	

22	km

23	

24	320	

25	5, 7, 2, 10	

TOTAL MARKS [ ]

D20 Name: .... Date: ....

1	°

2	kg

3	

4	cm

5	

6	%

7	79.6	

8	

9	

10	

11	£792	£

12	

13	511	

14	6000	

15	£450	£

16	

17	

18	$d = 2(e + 3)$	

19	

20	4 cm	cm²

21	11, 15p	p

22	17, 850	

23	896	

24	

25	$6x - 13 = 11$	

TOTAL MARKS [ ]

D21 Name: .... Date: ....

1 £	**11** litres	**21** 18, 720
2	**12** 216	**22** 12, 17
3 litres	**13** 904	**23** 897, 454
4	**14** $1\frac{2}{3}$ hrs — min	**24** 13
5 m	**15** km	**25** $y = 8x - 15$
6	**16** 5 : 8 — cm	
7 3.72	**17** $y = 5x^2$	
8 %	**18** %	
9	**19**	
10	**20** 16 — cm^2	**TOTAL MARKS**

- ✂ - - - -

D22 Name: .... Date: ....

1 km	**11** cm — A ———————— B	**20** £44 — £
2 °	**12** £450 — £	**21** 6.5, 14.5
3	**13** 721	**22** 840
4	**14** litres	**23** 684
5 mm	**15** 416	**24** 3, 6, 5, 7, 4, 5
6 9.89	**16**	**25**
7 %	**17** 70°, 80° — °	
8	**18** $a = c^2 + 5$	
9 m^2	**19**	
10 %		**TOTAL MARKS**

D23 Name: .... Date: ....

1	kg

2	kg

3	

4	°

5	

6	0.45	

7	%

8	0.96	

9	

10	

11	70	

12	403	

13	20%, 80	

14	

15	$3\frac{2}{3}$ hrs	min

16	30 ml	ml

17	740	£

18	$a = 6c^2$	

19	32 m	m²

20	40	

21	km

22	360	

23	7.5, 11.5	

24	

25	$7x - 12 = 51$	

TOTAL MARKS

D24 Name: .... Date: ....

1	m

2	

3	ml

4	

5	

6	0.08	

7	

8	cm²

9	%

10	

11	470 ft	ft

12	

13	min

14	698	

15	A ———————————— B

16	160	

17	$p = \dfrac{18}{q + 3}$	

18	4 : 5	£

19	20 m	m²

20	

21	487, 945	

22	18, 900	

23	66, 11	

24	17	

25	$y = 5x - 12$	

TOTAL MARKS

1	km

2	

3	

4	

5	

6	5.24	

7	

8	m²

9	

10	

11	417	£

12	

13	303	

14	m

15	°

16	100°, 60°, 140°	°

17	

18	340	

19	$d = 3(e - 4)$	

20	40	

21	240	

22	34, 11	

23	588	

24	1, 4, 5, 6, 4	

25	$5x - 11 = 19$	

TOTAL MARKS

1	°

2	cm

3	

4	kg

5	

6	4.11	

7	%

8	

9	

10	%

11	

12	$1\frac{1}{3}$ hrs	min

13	10%, 30	

14	£55	£

15	

16	$y = x^2 - 3$	

17	170	

18	

19	60°, 100°	°

20	%

21	5.5, 13.5	

22	340	

23	5, 3, 8, 5, 4, 5	

24	675, 487	

25	

TOTAL MARKS

D27 Name: .... Date: ....

1 ____ m²	**11** ____ cm A _____ B	**20** ____
2 ____	**12** 4.5 ____	**21** 15, £23 ____ £
3 ____	**13** 694 ____	**22** 240 ____
4 ____ cm	**14** ____ hrs	**23** ____ km
5 18 litres ____ gal	**15** 411 ____	
6 1.57 ____	**16** 21 cm² ____ cm	**24** 11 ____
7 ____	**17** $p = 4q^2$ ____	**25** ____
8 ____ litres	**18** 9 : 4 ____ km/hr	
9 ____	**19** ____	
10 ____		**TOTAL MARKS** ____

For item 23: triangle with vertices labelled C (top), B (right), A (bottom left).

D28 Name: .... Date: ....

1 ____ °	**11** ____	**20** ____ kg
2 ____ mm	**12** 302 ____	**21** 7.5, 13.5 ____
3 ____	**13** ____ °	**22** 40, 680 ____
4 ____ km	**14** 30 min ____ min	**23** 12, 8, 4, 3, 7, 2 ____
5 ____	**15** 217 ____	**24** ____
6 ____	**16** 420 ____	**25** 357, 955 ____
7 ____	**17** ____	
8 2.99 ____	**18** $s = 30 - 5t^2$ ____	
9 ____ %	**19** 130°, 120°, 80° ____ °	
10 ____ %		**TOTAL MARKS** ____

For item 13: an angle diagram.

D29 Name: .... Date: ....

1			11	140		21	24	
	m^2							

2			12	808		22	224, 14	

3			13	418		23	48, 11	
	g							

4			14	45		24	784	
					g			

5			15	$3\frac{3}{4}$ hrs		25	$y = 8x - 14$	
					min			

6	0.789		16		

7			17	£240	
	%				£

8			18	32 cm	
	m				cm^2

9			19	$d = (e - 3)^2$	

10			20		

TOTAL MARKS

D30 Name: .... Date: ....

1			11	130°, 120°, 80°	°	20	1.5	
	litres							

2			12	708		21		km

3			13		
	m				

4			14	324	
	°				

5	80 km		15			22	35, 12	
		miles		A ⊢———————⊣ B				

6						23	16, 960	

7			16	5 : 3		24	3, 5, 4, 9, 4	
					cm			

8			17			25		
	%							

9	0.712		18	$y = x^2 - 2$	

10			19		
	%				

TOTAL MARKS

D31 Name: .... Date: ....

#			#			#		
1	m²		**11**	520		**20**		
2			**12**	750		**21**	£976, £555	£
3			**13**			**22**	15, 450	
4	ml		**14**	193		**23**	12, 16	
5	°		**15**	mm				
6	cm			A _____ B		**24**		
7			**16**	£57	£			
8	%		**17**	$y = 13 - 2x$		**25**	$5x - 13 = 7$	
9	3.97		**18**					
10	0.16		**19**	28 mm	mm²		**TOTAL MARKS**	

- ✂ - - - - - - - -

D32 Name: .... Date: ....

#			#			#		
1	kg		**11**	587		**21**	868 days	wks
2			**12**	$1\frac{1}{3}$ hrs	min	**22**	3.5, 7.5	
3	yrs		**13**	203		**23**	19, 570	
4	cm		**14**	56		**24**	24	
5			**15**			**25**	$y = 7x - 13$	
6			**16**	$p = (q + 2)^2$				
7	97.8		**17**	160	£			
8	%		**18**	kg				
9	cm²		**19**	15 m²	m			
10			**20**				**TOTAL MARKS**	

D33 Name: .... Date: ....

1 £		**11** 177 kg → kg		**20** 85°, 80° → °
2		**12** → kg		**21** 6.5, 16.5
3		**13** 511		**22** 420
4 mm		**14** 0°		**23** 5, 6, 2, 4, 3
5 km				**24** 887, 659
6		**15**		**25**
7		**16** 5 : 1 → kg		
8 2.27		**17** 20		
9 %		**18** $c = 2(d + 3)$		
10		**19** %		**TOTAL MARKS** [ ]

---✂--------

D34 Name: .... Date: ....

1 km		**11** litres		**20** 90°, 130°, 70° → °
2		**12** 309		**21** £386, £735 → £
3 °		**13**		**22** 12, 15
4 ml		**14**		**23**
5 20 miles → km		**15** mm		**24** 840
6 m²		A _____ B		**25** km
7 %		**16**		
8 3.091		**17** $a = 2c^2$		
9		**18**		
10 %		**19** 70		**TOTAL MARKS** [ ]

#			#			#		
1	°		11	£811	£	20	64	
2			12	4000		21		
3	km		13	100		22	330	
4			14			23	776	
5	cm		15	A————————B		24	44, 11	
6	%		16	kg		25	$7x + 9 = 65$	
7	0.915		17	$y = x^2 - 4$				
8			18					
9	%		19	34 mm	mm²			
10								

TOTAL MARKS

#			#			#		
1	kg		11	km		21	645, 875	
2	°		12	210		22	720	
3	cm		13	312		23	5, 1, 9, 4, 6	
4			14	$2\frac{3}{4}$ hrs	min	24	44, 11	
5			15	483		25		
6			16					
7	m²		17	70°, 100°, 110°	°			
8	%		18	litres				
9	8.81		19	6 : 5	£			
10	0.16		20	$a = \dfrac{24}{c + 3}$				

TOTAL MARKS

1	10 gallons	litres

2	litres

3	

4	

5	

6	%

7	4.83	

8	

9	

10	

11	

12	g

13	221	

14	412	

15	A ————————————— B

16	%

17	24 m	m²

18	$d = 4(e - 2)$	

19	2.5	

20	56	

21	5.5, 11.5	

22	440	

23	6, 5, 2, 7	

24	876	

25	

TOTAL MARKS

- ✂ - - -

1	m²

2	

3	

4	°

5	m

6	0.095	

7	

8	cm

9	%

10	0.12	

11	400 cal	cal

12	70	

13	393	

14	

15	(diagram) 0°

16	5 : 7	mm

17	$y = 6x^2$	

18	

19	80°	°

20	

21	390	

22	5.5, 7.5	

23	

24	£896, £446	£

25	21	

TOTAL MARKS

Name: .... **Date:** ....

1	m
2	
3	
4	g
5	
6	
7	cm²
8	2.27
9	%
10	

11	£287 £
12	950
13	
14	g
15	A ———————— B
16	8100
17	£64 £
18	$d = (e + 3)^2$
19	

20	24 cm² cm
21	km

A D C B (diagram)

22	31, 12
23	480
24	
25	$y = 6x + 9$

TOTAL MARKS

Name: .... **Date:** ....

1	°
2	
3	
4	g
5	10 litres pints
6	%
7	0.672
8	
9	
10	

11	$4000 $
12	$2\frac{2}{3}$ hrs min
13	907
14	
15	511
16	90°, 110°, 90° °
17	
18	$y = x^2 + 3$
19	$
20	5 : 4 £

21	£864, 9 £
22	1, 5, 8, 10, 11
23	15, 11
24	550
25	$y = 9x - 14$

TOTAL MARKS

E1   Name: ....   Date: ....

1		cm

2	

3	

4	

5	

6	%

7	

8	

9	

10	

11	16 cm cm²	

12	%	

13	2200	

14	C = 25 – 3x	

15	£40 £	

16	985, 37	

17		

18	60 km/hr hrs 135 km min	

19	3x < 29	

20		

21	y = 5x + 9	

22		

23	14	

24		

25	$\dfrac{6.27 \times 4.93}{2.1}$	

TOTAL MARKS [ ]

E2   Name: ....   Date: ....

1		cm²

2	

3	

4	%

5	

6	

7	cm

8	0.002	

9	

10	$\dfrac{600}{1000}$	

11	65° °	

12	5 : 3, 60	

13	64	

14	$d = \dfrac{16}{e + 3}$	

15		

16	41, 49	

17	x + 3 ≥ 7	

18		

19	32 × 21 = 672	

20	260 km hrs 80 km/hr min	

21	y = 7x + 9	

22		

23	8x – 15 = 17	

24		

25	$\dfrac{3.7 \times 9.2}{1.8}$	

TOTAL MARKS [ ]

E3 Name: .... Date: ....

1		cm²
2		%
3		
4		
5		
6		mm
7		
8		
9		
10	0.08	

11	$p = (q + 3)^2$	
12		
13	30	
14	70°, 80°, 90°	°
15	7500	
16	860	
17	9%, 351	
18		
19	30 km, 20 min	km/hr
20	$x - 2 \geq 5$	

21	$y = 6x - 11$	
22	23	
23		
24	$\dfrac{40.2 \times 20.2}{1.9}$	
25		

TOTAL MARKS [ ]

--------------------------------✂--------------------------------

E4 Name: .... Date: ....

1	0.097	
2		
3		%
4		
5		%
6		
7	$\dfrac{200}{10000}$	
8		m
9		
10	0.03	

11	40 m², 4 m	m
12	490	
13	20	
14	$y = 3 (x - 2)$	
15	£63	£
16	870	
17		
18	381, 221	
19	$x^2 \geq 60$	
20	30 m/s, 6.5sec	m

21	$7x - 15 = 27$	
22		
23	$y = 6x - 14$	
24		
25	$\dfrac{19.4 \times 88.3}{4.2 \times 4.7}$	

TOTAL MARKS [ ]

E5 **Name:** .... **Date:** ....

1	cm²

11	60 cm	cm

21	

2	

12	4300	

22	4, 7, 6, 8	

3	%

13	$d = 3e^2$	

23	$21 = 37 - 4x$	

4	

14	90°, 70°	°

24	48 m/min	m/s

5	

15	2 : 1	ml

25	$\dfrac{76}{23 \times 17}$	

6	m

16	6 km, 40 min	km/hr

7	

17	

8	

18	$2x < 30$	

9	0.003	

19	

10	

20	51%, 201	

TOTAL MARKS

--✂----------

E6 **Name:** .... **Date:** ....

1	m

11	80°, 80°, 70°	°

21	78%, 297	

2	

12	48	

22	

3	%

13	

23	$y = 9x - 14$	

4	

14	£60	£

24	

5	

15	$a = \dfrac{12}{c + 2}$	

25	$\dfrac{24.3 \times 8.22}{9.7 \times 1.3}$	

6	0.6	

16	281, 57	

7	

17	$2x - 1 \leq 14$	

8	

18	92	

9	$\dfrac{400}{1000}$	

19	

10	0.002	

20	100 km hrs 30 km/hr min

TOTAL MARKS

E7 Name: .... Date: ....

1		g

11	16 cm	cm²

21	

2	

12	620	

22	2, 7, 9, 6	

3	

13	

23	$7x - 13 = 36$	

4	%

14	$y = x^2 + 4$	

24	

5	

15	58	

25	$\dfrac{49 \times 52}{47 \times 11}$	

6	

16	

7	cm²

17	$x - 2 > 2$	

8	0.05	

18	90 km/hr $3\frac{2}{3}$ hrs	km

9	$\dfrac{30}{10000}$	

19	504, 1504	

10	

20	

TOTAL MARKS [ ]

--✂------------

E8 Name: .... Date: ....

1	cm²

11	60°, 80°	°

21	$\dfrac{8.63 \times 105}{48.7}$	

2	

12	60%	

22	$y = 44 - 6x$	

3	

13	$p = 20 - 4q$	

23	

4	%

14	2400	

24	3, 5, 4, 9, 4	

5	%

15	10 : 3, 40 cm	cm

25	

6	

16	39%, £397	£

7	

17	$x + 2 \geq 3$	

8	

18	

9	0.42	litres

19	3 hrs 20 min 390 km	km/hr

10	0.04	

20	0.08 cm	cm

TOTAL MARKS [ ]

E9 Name: .... Date: ....

1	kg		**11**	$d = 2(e + 3)$		**21**	$\dfrac{19.7 \times 19}{5 \times 41}$	
2	9.505		**12**	%		**22**	91%, 703	
3			**13**	60p	p	**23**	$6x + 9 = 57$	
4			**14**	28 cm, 4 cm	cm²	**24**		
5	%		**15**			**25**		
6			**16**	165 km 60 km/hr	hrs min			
7	%		**17**					
8	$\dfrac{300}{1000}$		**18**	903, 49				
9			**19**	$2x < 11$				
10			**20**			**TOTAL MARKS**		

--✂------

E10 Name: .... Date: ....

1	cm²		**11**	16		**21**	$\dfrac{19 \times 63}{1.1 \times 2.9}$	
2	%		**12**	660		**22**	8, 12, 8, 16	
3			**13**	88 mm	mm	**23**	$y = 5x - 11$	
4			**14**	100°, 90°, 100°	°	**24**		
5			**15**	$y = \dfrac{20}{x + 2}$		**25**	0.3 m/s	m/min
6	0.003		**16**	£86	£			
7	m		**17**	80 km/hr 2 hrs 45 min	km			
8			**18**	57, 72				
9			**19**	$2x + 1 \geq 6$				
10			**20**	$73 \times 37 = 2701$		**TOTAL MARKS**		

E11 Name: .... Date: ....

1		cm

11	2 : 1	min

21	11	

2		

12	4100	

22		

3		

13	$y = 5x^2$	

23	$8x - 16 = 24$	

4		%

14	20 mm	mm²

24		

5		

15	£20	£

25	$\dfrac{372 \times 22.9}{214 \times 1.9}$	

6		

16		

7		

17	0.07 g	g

8		

18	71%, 405	

9	0.03	

19	200 km, $2\frac{1}{2}$ hrs	km/hr

10		

20	$x - 1 > 4$	

TOTAL MARKS [ ]

---✂------------

E12 Name: .... Date: ....

1		cm²

11	65°, 85°	°

21	$y = 5x - 13$	

2		%

12	300	

22		

3		

13	$a = c^2 + 5$	

23	3, 11, 9, 5	

4		

14	180	

24	$\dfrac{43 \times 48}{9.7 \times 2.1}$	

5		%

15	$45	$

25		

6		kg

16		

7	0.003	

17	491, 47	

8	$\dfrac{2000}{10000}$	

18	315 km, 70 km/hr	hrs min

9	0.007	

19	$x + 1 < 8$	

10		

20		

TOTAL MARKS [ ]

E13   Name: ....   Date: ....

1		litres

11	88	mm

21	$7 = 49 - 7x$	

2	

12	$a = 6c^2$	

22	

3	

13	

23	3, 5, 4, 5, 8, 5	

4	

14	30	cm

24	$\dfrac{1.1 \times 4.8}{5.1}$	

5	

15		%

25	

6	

16	70 km/hr, $3\frac{1}{2}$ hrs	km

7		%

17	$10x < 15$	

8	

18	

9	

19	

10	

20	51, 81	

TOTAL MARKS [ ]

--✂-----

E14   Name: ....   Date: ....

1	0.08	

11	4 : 1	kg

21	69%, £598	£

2		%

12	380	

22	

3	

13	700	

23	$7x - 14 = 14$	

4		cm²

14	80°, 130°, 70°	°

24	$\dfrac{99.3}{1.12 \times 4.69}$	

5	

15	$p = \dfrac{18}{q+3}$	

25	54 km/hr	km/min

6		mg

16	630	

7	

17	

8	$\dfrac{5000}{10000}$	

18	8 km/hr, $2\frac{1}{2}$ hrs	km

9	

19	83, 409	

10	0.0007	

20	$x - 10 > 1$	

TOTAL MARKS [ ]

E15 Name: .... Date: ....

1	0.003		**11**	75°, 90°	°	**21**	
2	%		**12**	170		**22**	
3			**13**	$d = 3(e - 4)$		**23**	$y = 5x - 15$	
4			**14**	3300		**24**	6, 5, 2, 7	
5			**15**			**25**	$\dfrac{19.3 \times 49.4}{4.9 \times 2.1}$	
6			**16**				
7	cm		**17**	695, 349			
8			**18**	175 km, $2\frac{1}{2}$ hrs	km/hr		
9			**19**	$2x + 1 \geq 27$			
10			**20**				

TOTAL MARKS [ ]

--------------------------------✂--------------------------------

E16 Name: .... Date: ....

1	cm²		**11**	40 m	m²	**21**	24	
2	%		**12**	180°	°	**22**	$y = 38 - 6x$	
3			**13**	$y = x^2 - 3$		**23**	
4			**14**			**24**	
5			**15**	68		**25**	$\dfrac{11.8 \times 98.3}{214}$	
6	$\dfrac{70}{10000}$		**16**	11%, £451	£		
7	ml		**17**				
8			**18**	$3x < 25$			
9			**19**				
10			**20**	90 km/hr $4\frac{1}{2}$ hrs	km		

TOTAL MARKS [ ]

1		tonne

11	130°, 70°, 80°	°

21	9x – 14 = 58	

2	

12	5300	

22	

3	

13	

23	y = 8x – 18	

4	%

14	p = 4q²	

24	

5	

15	120, 5 : 7	

25	$\frac{239 \times 799}{10.3 \times 2.4}$	

6	

16	

7	

17	208, 44	

8	

18	480	

9	

19	2x + 1 ≤ 35	

10	

20	210 km 90 km/hr	hrs min

TOTAL MARKS

1	cm²

11	120	

21	42 – 4x = 10	

2	%

12	18 mm	mm²

22	3, 6, 5, 7, 4, 5	

3	

13	s = 30 – 5t	

23	

4	

14	830	

24	

5	

15	63	litres

25	$\frac{20.4 \times 3.7}{1.9 \times 1.8}$	

6	$\frac{90}{10000}$	

16	41, 58	

7	kg

17	

8	0.007	

18	

9	

19	x² < 16	

10	

20	25 km, 2½ hrs	km/hr

TOTAL MARKS

E19 Name: .... Date: ....

1		m

| 2 | | % |

| 3 | |

| 4 | |

| 5 | |

| 6 | |

| 7 | |

| 8 | |

| 9 | |

| 10 | 0.0002 | |

11	70°	°
12	670	
13	$d = (e - 3)^2$	
14		%
15	35 m	m
16	62, 78	
17		
18	$2x + 1 \geq 15$	
19		
20	30 m/s, 135m	sec

21	89°, £396	£
22	$y = 6x - 13$	
23		
24		
25	$\dfrac{3.37 \times 1.98}{6.12}$	

TOTAL MARKS

E20 Name: .... Date: ....

1		km
2		
3		
4		
5		%
6		%
7	$\dfrac{8000}{10000}$	
8		
9		
10		

11	5 : 3, 36	
12	$y = x^2 - 2$	
13		
14	£4.00	£
15	12 m	m²
16	940	
17	30 m/s, $3\frac{1}{2}$ sec	m
18	594, 29	
19	$41 \times 38 = 1558$	
20	$4x \leq 17$	

21	$5x - 12 = 33$	
22		
23		
24	$y = 6x - 12$	
25	$\dfrac{50.4 \times 11.1}{5.4}$	

TOTAL MARKS

E21 Name: .... Date: ....

1		cm²

11	90°, 60°, 140°	°

21	0.7 km/min	km/hr

2		%

12	55	

22	12	

3	

13	570	$

23	$8x - 14 = 34$	

4	

14	$y = 13 - 2x$	

24	

5	

15	

25	$\dfrac{27 \times 11}{29 \times 19}$	

6	

16	$x + 4 < 13$	

7		m

17	

8	

18	

9	

19	62%, £904	£

10	

20	150 km 2 hrs 30 min	km/hr

TOTAL MARKS

E22 Name: .... Date: ....

1		°C

11	80	litres

21	$y = 38 - 5x$	

2	

12	

22	

3		%

13	$C = 2 (d + 3)$	

23	5, 7, 2, 10	

4	

14		%

24	

5		%

15	28 m²	m

25	$\dfrac{99 \times 8}{1.3 \times 1.6}$	

6	

16	880	

7	

17	42, 77	

8	$\dfrac{20}{10000}$	

18	

9	0.001	

19	$3x > 30$	

10	

20	90 km/hr 2 hrs 40 min	km

TOTAL MARKS

E23   Name: ....   **Date:** ....

1	cm²
2	
3	
4	%
5	
6	m
7	0.004
8	
9	$\frac{4000}{10000}$
10	

11	£160, 5 : 3 £
12	7300
13	$p = (q + 2)^2$
14	75°, 85° °
15	kg
16	$5x < 24$
17	
18	204, 804
19	4 km, 40 min km/hr
20	

21	$5x + 9 = 24$
22	5, 6, 2, 4, 3
23	
24	
25	$\frac{13.7}{1.2 \times 2.2}$

TOTAL MARKS

E24   Name: ....   **Date:** ....

1	ml
2	%
3	cm²
4	
5	
6	
7	$\frac{6000}{10000}$
8	
9	0.001
10	

11	28 m mm²
12	
13	
14	$a = 2c^2$
15	
16	57, 32
17	
18	30 km/hr hrs 80 km min
19	$2x - 1 \leq 18$
20	

21	
22	
23	$y = 7x - 13$
24	49%, 195
25	$\frac{11.3 \times 46.2}{1.9 \times 5.4}$

TOTAL MARKS

Name: .... Date: ....

1	m/s		**11**	100°, 70°, 100°	°	**21**	$\dfrac{13.3 \times 1.13}{1.4}$	

1   m/s

11   100°, 70°, 100°   °

21 $\dfrac{13.3 \times 1.13}{1.4}$

2

12 $y = x^2 - 4$

22 $6x - 12 = 30$

3

13   %

23

4

14

24 $y = 5x - 13$

5

15   680

25   10 m/s   km/hr

6   0.002

16   60 km/hr $3\frac{1}{2}$ hrs   km

7   %

17   77,   204

8

18 $x^2 < 121$

9

19

10   0.009

20 $59 \times 19 = 1121$

TOTAL MARKS

Name: .... Date: ....

1   °

11   100°, 70°, 100°   °

21 $\dfrac{297}{101 \times 1.05}$

2   %

12   6 : 13

22   2, 9, 3, 8, 5, 3

3

13   g

23 $40 - 6x = 10$

4

14   6100

24

5

15 $a = \dfrac{24}{c + 3}$

25

6   0.001

16

7

17   590

8

18   38,   79

9 $\dfrac{50}{10000}$

19 $x - 4 \geq 72$

10

20   6 km/hr   hrs   21 km   min

TOTAL MARKS

© Graham Newman 1997. Published by Thomas Nelson and Sons Ltd. Copyright permitted for purchasing schools only.

E27   Name: ....   Date: ....

1	cm²

2	%

3	

4	

5	

6	0.03	

7	0.42	litres

8	

9	

10	

11	460	

12	$d = 4 (e - 2)$	

13	6300	

14	35	

15	8 cm	cm²

16	

17	22%, 304	

18	

19	$x + 2 \leq 23$	

20	315 km, $3\frac{1}{2}$ hrs	km/h

21	33	

22	

23	$y = 8x - 12$	

24	

25	$\dfrac{28.7 \times 4.99}{4.6 \times 3.2}$	

TOTAL MARKS

E28   Name: ....   Date: ....

1	m

2	

3	

4	

5	

6	%

7	

8	

9	$\dfrac{60}{10000}$	

10	

11	

12	3500	

13	$y = 6x^2$	

14	60°, 70°	°

15	

16	

17	

18	297, 31	

19	80 km/hr $3\frac{1}{2}$ hrs	km

20	$x + 2 \geq 3$	

21	$7x - 13 = 50$	

22	6, 1, 4, 5, 4	

23	

24	

25	$\dfrac{71.4 \times 49.3}{4.9 \times 6.9}$	

TOTAL MARKS

E29 Name: .... Date: ....

1	cm²

| 2 | % |

| 3 | |

| 4 | |

| 5 | |

| 6 | |

| 7 | °F |

| 8 | $\dfrac{800}{1000}$ | |

| 9 | |

| 10 | |

11	26 mm	mm²

| 12 | |

| 13 | |

| 14 | $d = (e + 3)^2$ | |

| 15 | 250g, 12 : 13 | g |

| 16 | 730 | |

| 17 | $x - 1 < 6$ | |

| 18 | |

| 19 | 25, 82 | |

| 20 | 135 km, $4\frac{1}{2}$ hrs | km/hr |

21	$\dfrac{2.4 \times 6.1}{1.4 \times 1.8}$	

| 22 | 69%, £598 | £ |

| 23 | $y = 7x - 12$ | |

| 24 | |

| 25 | 360 km/hr | m/s |

TOTAL MARKS [ ]

--✂--------

E30 Name: .... Date: ....

1	km

| 2 | % |

| 3 | |

| 4 | |

| 5 | % |

| 6 | 0.004 | |

| 7 | |

| 8 | |

| 9 | $\dfrac{900}{1000}$ | |

| 10 | |

11	110°, 130°, 70°	°

| 12 | $y = x^2 + 3$ | |

| 13 | |

| 14 | % |

| 15 | 580 | |

| 16 | $x^2 < 81$ | |

| 17 | 12 km/hr · 28 km | hrs · min |

| 18 | 0.009 | |

| 19 | $16 \times 36 = 576$ | |

| 20 | 749, 821 | |

21	$\dfrac{24.2 \times 43.7}{1.2 \times 2.2}$	

| 22 | |

| 23 | $6x - 11 = 37$ | |

| 24 | |

| 25 | 15 | |

TOTAL MARKS [ ]

E31 Name: .... Date: ....

1	m²		11	£80	£		21	$\frac{31.5 \times 2.2}{3.6 \times 4.9}$	
2			12	16 mm	mm²		22	$y = 45 - 4x$	
3			13	360			23	$5x - 16 = 19$	
4	%		14	$C = 25 - 3x$			24		
5			15	99			25		
6	°		16	30 m/s, $2\frac{1}{2}$sec	m				
7	0.002		17						
8			18	9%, £748	£				
9			19	$x + 1 \leq 25$					
10			20						

TOTAL MARKS [ ]

- ✂ - - - - - -

E32 Name: .... Date: ....

1	amps		11	2300			21	$\frac{270 \times 140}{18 \times 12}$	
2			12	$p = (q + 3)^2$			22	82%, 802	
3	%		13				23	$y = 8x - 12$	
4	cm²		14	65°, 75°	°		24		
5			15	10 : 7, 49			25	8 litres/s	litres/min
6			16	4 km, 40 min	km/hr				
7			17	709, 96					
8	0.03		18						
9			19	$4x > 19$					
10	0.008		20						

TOTAL MARKS [ ]

E33   Name: .... Date: ....

1	°C

2	

3	

4	

5	

6	%

7	0.002	

8	

9	$\dfrac{9000}{10000}$	

10	0.04	

11	$16	$

12	$y = 3(x - 2)$	

13	820	

14	%

15	10 cm²	cm

16	635, 32	

17	940	

18	

19	$x^2 \leq 50$	

20	60 km/hr 3 hrs 45 min	km

21	$\dfrac{8.6 \times 32}{6.4 \times 11}$	

22	5, 3, 9, 3	

23	$6x - 13 = 11$	

24	

25	

TOTAL MARKS

E34   Name: .... Date: ....

1	sec

2	%

3	

4	cm²

5	

6	

7	

8	

9	

10	

11	24 m	m²

12	%

13	450	

14	m

15	$y = \dfrac{20}{x + 2}$	

16	

17	

18	80 km/hr 300 km	hrs min

19	18%, 697	

20	$x + 6 > 24$	

21	$y = 42 - 4x$	

22	

23	$7x - 12 = 51$	

24	$\dfrac{42.1 \times 20.4}{1.9}$	

25	

TOTAL MARKS

1	°		**11**	100°, 110°, 90°	°	**21**	$\dfrac{760 \times 113}{10 \times 2.2}$
2			**12**	£96, 5 : 3	£	**22**	$y = 8x - 15$
3			**13**	$d = 3e^2$		**23**	
4			**14**	390		**24**	
5	%		**15**	£		**25**	14
6			**16**				
7	0.02		**17**	$4x > 22$			
8	$\dfrac{200}{1000}$		**18**	40 min, 20 km	km/hr		
9	%		**19**	298, 808			
10	0.006		**20**	$36 \times 49 = 1764$			

TOTAL MARKS

1	cm²		**11**	32 mm²	mm	**21**	$\dfrac{42 \times 22}{19 \times 12}$
2			**12**			**22**	7, 4, 2, 5, 2
3			**13**	$y = x^2 + 4$		**23**	$41 - 5x = 11$
4	%		**14**			**24**	
5			**15**			**25**	
6	cm		**16**	470			
7	0.001		**17**	474, 53			
8			**18**				
9	$\dfrac{80}{10000}$		**19**	$x + 2 \geq 16$			
10	0.004		**20**	90 km/hr 3 hrs 20 min	km		

TOTAL MARKS

E37 Name: .... Date: ....

1	°F	

| **2** | % | |

| **3** | | |

| **4** | | |

| **5** | % | |

| **6** | | |

| **7** | | |

| **8** | | |

| **9** | | |

| **10** | 0.002 | |

| **11** | 1 : 2 | |

| **12** | $p = 20 - 4q$ | |

| **13** | 8400 | |

| **14** | 60° | ° |

| **15** | | $ |

| **16** | 719, 52 | |

| **17** | 60 km, 40 min | km/hr |

| **18** | | |

| **19** | $x^2 < 144$ | |

| **20** | | |

| **21** | $y = 5x - 12$ | |

| **22** | | |

| **23** | 58%, 799 | |

| **24** | $\dfrac{82 \times 195}{42 \times 2.1}$ | |

| **25** | | |

TOTAL MARKS

✂

E38 Name: .... Date: ....

| **1** | kg | |

| **2** | % | |

| **3** | | |

| **4** | | |

| **5** | | |

| **6** | | |

| **7** | 0.003 | |

| **8** | | |

| **9** | | |

| **10** | 0.003 | |

| **11** | 300 | |

| **12** | 370 | |

| **13** | $d = 2 (e + 3)$ | |

| **14** | 4 cm | cm² |

| **15** | 50 | litres |

| **16** | 630 | |

| **17** | 12 km/hr 9 km | hrs min |

| **18** | | |

| **19** | 43, 83 | |

| **20** | $10x \geq 31$ | |

| **21** | $\dfrac{48 \times 39}{0.9 \times 4.2}$ | |

| **22** | $y = 8x - 14$ | |

| **23** | $5x - 11 = 19$ | |

| **24** | | |

| **25** | | |

TOTAL MARKS

E39 Name: .... Date: ....

1	0.19	litres	**11**	80°, 70°, 80°	°	**21** $\dfrac{36 \times 102}{9.6 \times 1.8}$
2	%		**12**	%		**22** 12, 4, 8, 3, 7, 2
3			**13** $y = 5x^2$			**23** $y = 36 - 3x$
4			**14**			**24**
5			**15** 20 m	m		**25** 18 litres/min litres/sec
6			**16** 5 hrs, 70 km/hr	km		
7			**17**			
8			**18** 304, 404			
9 $\dfrac{700}{1000}$			**19** $x + 2 \geq 21$			
10			**20**			**TOTAL MARKS**

---✂--------

E40 Name: .... Date: ....

1			**11** 4 out of 10			**21** 42%, 401
2			**12** $a = c^2 + 5$			**22**
3			**13** 9200			**23** $5x - 13 = 7$
4	%		**14** $3000	$		**24**
5	cm²		**15** 18	cm²		**25** $\dfrac{39}{1.92 \times 2.1}$
6			**16** 212, 386			
7	ml		**17**			
8	0.09		**18** $4x \leq 23$			
9 $\dfrac{40}{10000}$			**19** 980			
10			**20** 3 km, 20 min	km/hr		**TOTAL MARKS**